Building Modern SaaS Applications with C# and .NET

Build, deploy, and maintain professional SaaS applications

Andy Watt

BIRMINGHAM—MUMBAI

Building Modern SaaS Applications with C# and .NET

Group Product Manager: Gebin George

Publishing Product Manager: Kunal Sawant

Senior Editor: Kinnari Chohan

Technical Editor: Jubit Pincy

Copy Editor: Safis Editing

Project Manager: Prajakta Naik

Proofreader: Safis Editing

Indexer: Hemangini Bari

Production Designer: Alishon Mendonca

Developer Relations Marketing Executive: Sonia Chauhan

First published: June 2023

Production reference: 1240523

Published by Packt Publishing Ltd.

Livery Place

35 Livery Street

Birmingham

B3 2PB, UK.

ISBN 978-1-80461-087-9

www.packtpub.com

Contributors

About the author

Andy Watt started his career as a programmer aged about eight years old on a Commodore 64, and he has never looked back! Andy started his career writing web based applications with .NET 1.1, and has worked with every version released since. These days Andy works as a consultant on a range of tools and projects, each with a common thread running through them – .NET. Andy currently enjoys being self-employed and using his skills, gained over many years, to help customers develop their tools and deliver their projects.

Outside of programming, Andy enjoys rock climbing, cycling, and walking in the hills and mountains of Scotland.

About the reviewers

John Ek is a seasoned software engineer and has over 25 years' experience in the software industry. John is currently a senior software engineer with GovX.

Ahmed Ilyas has 18 years of professional experience in software development.

Previously working for Microsoft, he then ventured into setting up a consultancy company offering the best possible solutions for a host of industries and providing real-world answers to problems faces in those industries. He only uses the Microsoft stack to build these technologies and bring in the best practices, patterns, and software to his client base to enable long-term stability and compliance in the ever-changing software industry, as well as to improve software developers around the globe and push the limits of technology.

He has been awarded the MVP in C# title three times by Microsoft for "providing excellence and independent real-world solutions to problems that developers face." With the great reputation that he has, this has resulted in having a large client base for his consultancy company, Sandler Ltd (UK) and Sandler Software LLC (US), which includes clients from different industries. Clients have included him on their "approved contractors/consultants" lists as a trusted vendor and partner.

Ahmed has also been involved in reviewing books for Packt Publishing.

I would like to thank the author/publisher of this book for giving me the great honor and privilege of reviewing the book. I would also like to thank Microsoft Corporation and my colleagues for enabling me to become a reputable leader as a software developer and provide solutions to the industry.

Table of Contents

2

Building a Simple Demo Application   21

Part 2: Building the Backend

3

What Is Multi-Tenancy, and Why Is It Important in SaaS Applications?   53

4

Building Databases and Planning for Data-Rich Applications   83

5

Building Restful APIs   103

6

Microservices for SaaS Applications   135

Part 3: Building the Frontend

7

Building a User Interface   163

8

Authentication and Authorization   193

Part 4: Deploying and Maintaining the Application

9

Testing Strategies for SaaS Applications   217

10

Monitoring and Logging   237

11

Release Often, Release Early   253

12

Part 5: Concluding Thoughts

13

Preface

I have been working on .NET applications virtually from the very beginning of .NET. I think that I have worked with every possible .NET tool over the years, frontend and backend, good and bad! This has memorably included delivering training tools for processing facilities, building remote monitoring tools, and developing optimization solutions – all with .NET!

In this book, I have tied together my experience from the many projects that I have worked on over the last 20 years to put together what I hope is an interesting and informative look at how to build modern **software-as-a-service (SaaS)** applications with .NET.

This book is designed to detail and demystify the world of building SaaS applications with .NET and the Microsoft tech stack and aims to deliver the technical knowledge required to build, test, deploy, and maintain a modern SaaS application. We start off by introducing the technology that we will use and showcase it with a demo application. As we move through the book, we add additional tools and techniques, building up to the point that we have described and demonstrated all of the major topics required to build a SaaS application.

In reading this book, you'll learn the tools, techniques, and best practices required to develop modern SaaS applications that will impress your users and deliver value for your enterprise. We cover the full stack, from the database to the **user interface (UI)**, and then onward to deploying and maintaining the application in production.

I have tried to make this book accessible to everyone. To this end, I have exclusively used free and open source tools, and I have tried to always introduce new topics by linking them to core principles with as little assumed knowledge as possible. Where possible, I have illustrated the theory with practical examples.

By the end of this book, you will have a comprehensive understanding of the theory required and will have explored much of it through practical examples. You should be well on your way to delivering your first real-world SaaS application!

It is my hope that you will find this book not only educational and informative but also enjoyable. I also very much hope that the learning that you take away from this book will help you in your career and in all of your future endeavors!

Who this book is for

This book is aimed at anyone who wants to build a SaaS application and is interested in using the Microsoft tech stack to build it! However, there is no shying away from the fact that SaaS development is a relatively advanced topic. It will stand you in good stead if you have a strong knowledge of C# and .NET. It will also be very useful to have a working knowledge of database and web development.

While this book covers advanced topics, do not be put off if you are still a beginner! In this book, even the more advanced topics are introduced and explained in an approachable way and generally underpinned with examples, such that there will be something to learn for everyone who wants to learn – regardless of skill level!

What this book covers

Chapter 1, The Modern Distribution Model that We All Need, gives a short history of SaaS and introduces the technologies that we will use in this book.

Chapter 2, Building a Simple Demo Application, dives into the technology that we will focus on in this book and builds a skeleton application that we can add to as we go along.

Chapter 3, What Is Multi-Tenancy, and Why Is It Important in SaaS Applications?, covers multi-tenancy, with a specific focus on SaaS applications. We add multi-tenancy to the example from *Chapter 2*.

Chapter 4, Building Databases and Planning for Data-Rich Applications, focuses on the database, specifically building with SQL Server and Entity Framework.

Chapter 5, Building RESTful APIs, builds on what we have covered with the database layer by adding an API to make the data available over HTTP. As well as building the example application, this chapter delves into the theory behind RESTful APIs.

Chapter 6, Microservices for SaaS Applications, looks at microservices and SaaS, which go hand in hand. This chapter covers the theory and practice of building a microservice application and applies some of the learnings to the demo app.

Chapter 7, Building a User Interface, explains why the UI is a very important piece of the SaaS puzzle. This chapter covers the theory and adds a UI to the demo app with Blazor.

Chapter 8, Authentication and Authorization, covers the complex topic of adding authentication and authorization to a SaaS app. Specific mention is made of the challenges involved in working with multi-tenant and microservice applications.

Chapter 9, Testing Strategies for SaaS Applications, explains how testing is very important when building SaaS applications. In this chapter, we look at the how and why of testing, including unit testing, integration testing, and end-to-end testing.

Chapter 10, Monitoring and Logging, illustrates that without these two distinct but related concepts, identifying and resolving issues in production applications can be very challenging. This chapter covers both topics and gives advice on which tools and techniques to use.

Chapter 11, Release Often, Release Early, covers the process of building and releasing your application in a managed way using **continuous integration/continuous deployment (CI/CD)** pipelines.

Chapter 12, Growing Pains – Operating at Scale, discusses how as a SaaS application starts to grow and gain users, a whole new set of problems arise and must be faced by the development team. This chapter covers these challenges and gives instructions on how to resolve them.

Chapter 13, Wrapping It Up, revisits what we have covered in this book and takes a look at how the reader can deploy their newfound SaaS knowledge!

To get the most out of this book

This book is intended to be as accessible as possible for anyone wanting to get into building SaaS applications. That said, SaaS is a complex and advanced topic, and so a good knowledge of the following will help:

- .NET application development with C#

- SQL Server and Entity Framework

- General web development

- Some knowledge of Docker may be useful but is not required

I hope, though, that anyone working in the software industry or still learning their craft can read this book and advance their understanding!

The following tools and technologies are used throughout:

Software/hardware covered in the book	Operating system requirements
Visual Studio Code	Windows, macOS, or Linux
Docker Desktop	Windows, macOS, or Linux
.NET v7	Windows, macOS, or Linux
Entity Framework	Windows, macOS, or Linux
Blazor	Windows, macOS, or Linux
SQL Server	Windows, macOS, or Linux

All of the initial setup instructions are given in *Chapter 2*. Any further setup is detailed as required as you work through this book.

If you are using the digital version of this book, we advise you to type the code yourself or access the code from the book's GitHub repository (a link is available in the next section). Doing so will help you avoid any potential errors related to the copying and pasting of code.

Download the example code files

You can download the example code files for this book from GitHub at https://github.com/
PacktPublishing/Building-Modern-SaaS-Applications-with-C-and-.NET.
If there's an update to the code, it will be updated in the GitHub repository.

We also have other code bundles from our rich catalog of books and videos available at https://
github.com/PacktPublishing/. Check them out!

Download the color images

We also provide a PDF file that has color images of the screenshots and diagrams used in this book.
You can download it here: https://packt.link/IOZxh.

Conventions used

There are a number of text conventions used throughout this book.

Code in text: Indicates code words in text, database table names, folder names, filenames, file
extensions, pathnames, dummy URLs, user input, and Twitter handles. Here is an example: "The Up
method creates the table in the database, and the Down method drops the table. This is converted into
SQL code, which is issued to the database engine when the database update command is issued. "

A block of code is set as follows:

```
[HttpPut("{id}")]
public async Task<IActionResult> UpdateAsync(int id, UpdateHabitDto
request)
{
    var habit = await _habitService.UpdateById(id, request);
    if (habit == null)
    {
        return NotFound();
    }
    return Ok(habit);
}
```

> **Tips or important notes**
> Appear like this.

Bold: Indicates a new term, an important word, or words that you see onscreen. For instance, words
in menus or dialog boxes appear in **bold**. Here is an example: "Sara clicks on the **Add New** button
under the list of habits."

Get in touch

Feedback from our readers is always welcome.

General feedback: If you have questions about any aspect of this book, email us at `customercare@packtpub.com` and mention the book title in the subject of your message.

Errata: Although we have taken every care to ensure the accuracy of our content, mistakes do happen. If you have found a mistake in this book, we would be grateful if you would report this to us. Please visit `www.packtpub.com/support/errata` and fill in the form.

Piracy: If you come across any illegal copies of our works in any form on the internet, we would be grateful if you would provide us with the location address or website name. Please contact us at `copyright@packt.com` with a link to the material.

If you are interested in becoming an author: If there is a topic that you have expertise in and you are interested in either writing or contributing to a book, please visit `authors.packtpub.com`.

Share Your Thoughts

Once you've read *Building Modern SaaS Applications with C# and .Net*, we'd love to hear your thoughts! Scan the QR code below to go straight to the Amazon review page for this book and share your feedback.

https://packt.link/r/1-804-61087-9

Your review is important to us and the tech community and will help us make sure we're delivering excellent quality content.

Download a free PDF copy of this book

Thanks for purchasing this book!

Do you like to read on the go but are unable to carry your print books everywhere?

Is your eBook purchase not compatible with the device of your choice?

Don't worry, now with every Packt book you get a DRM-free PDF version of that book at no cost.

Read anywhere, any place, on any device. Search, copy, and paste code from your favorite technical books directly into your application.

The perks don't stop there, you can get exclusive access to discounts, newsletters, and great free content in your inbox daily

Follow these simple steps to get the benefits:

1. Scan the QR code or visit the link below

https://packt.link/free-ebook/9781804610879

2. Submit your proof of purchase
3. That's it! We'll send your free PDF and other benefits to your email directly

Part 1: Getting Started

This section starts by giving the history of **software-as-a-service** (**SaaS**) and introduces the demo application that we will build as we move through this book.

This section has the following chapters:

- *Chapter 1, The Modern Distribution Model That We All Need*
- *Chapter 2, Building a Simple Demo Application*

SaaS – the Modern Distribution Model that We All Need

This book will take you, the reader, through the full process of building a **software-as-a-service** (**SaaS**) application. But before we get our hands dirty, we'll start with an introduction to the broad concepts of SaaS. The history of application development is not particularly long, but it has been eventful! We will look at the last 40 years of software engineering to see how we have arrived at this point, with SaaS emerging as the dominant paradigm, and we will consider why SaaS has emerged as such. There are benefits to businesses and teams of all sizes in utilizing SaaS.

Finally, we will cover the tools and techniques that you as a developer will learn to enable you to confidently solve real-world problems by building applications using the SaaS paradigm. This chapter will form an introduction to the topic. We'll not deep dive into any specific skills (that will come later!). Rather, we will set the scene for the rest of this book, in which we will learn how to build, test, and release modern SaaS applications using the Microsoft tech stack.

This chapter covers the following main topics:

- What is SaaS?
- What other types of applications are there?
- Where did it all begin?
- Why is SaaS a popular choice for businesses large and small?
- Which tools are required to build SaaS apps?
- Which techniques will be used to build SaaS apps?
- How does this affect the development process?
- What to expect from this book
- How to assess and pay off technical debt

By the end of this chapter, you'll have learned the definition of SaaS. You'll have covered a brief history of application development, and you'll have learned the tools and techniques that we will see as we progress through this book!

What is SaaS?

SaaS has become the dominant paradigm in delivering an application to users. But what *is* SaaS?

You could say that SaaS applications are software applications that are delivered to their users over the internet and in a browser, often paid for by way of a monthly subscription fee.

And while that is technically correct, that single-sentence definition glosses over a *lot* of complexity! A lot of complexity for the user, the vendor, and most certainly for you—the developer of such applications!

In this chapter, we will start to build an understanding of SaaS, with a basic understanding of the technical, operational, and functional knowledge required to build a SaaS application from the ground up using Microsoft technologies. We will expand on that previous definition to the point that you—the reader, and soon-to-be SaaS developer—can confidently approach and solve problems by delivering SaaS-based solutions!

Let's start by expanding on that definition a little.

SaaS is typically understood to be applications that are delivered and interacted with through a browser. The software is not purchased or "owned" by the user and installed on their computer. Instead (typically), a monthly *membership* fee is paid, which grants the user access to the *service*.

From the user's point of view, the benefit of this is that they can access the application anywhere, at any time, on any device. They do not have to worry about installing the application or keeping the app up to date. All of that is *just part of the service*.

Many of the biggest tech companies in the world provide SaaS applications, and there is a good chance that you are using at least one of them!

Gmail is a great example of a SaaS application that is provided by Google. While it is free to most users, there is a fee for corporate or professional users who must pay for access to the G Suite tools. As well as a SaaS mail client, Google also offers a calendar application and a contact management application, as well as Office tools for authoring documents, building spreadsheets, or creating slides for a presentation.

For a small monthly fee, you can use your browser to access the Shopify application, which is—of course—a SaaS application.

All social media sites are also examples of SaaS applications, and while they may be free to use for the vast majority of the user base, there is a cost to the businesses that use the platforms to advertise their products.

As well as the examples given previously featuring enormous, multinational, billion-dollar corporations, there are many more examples of businesses that are supplying their software using the SaaS paradigm. There are software vendors—and from the vendors' point of view, there are many benefits to delivering software in this way. The primary and most obvious benefit of this is that they have a huge market reach—a truly global market that they can tap into, with multiple very high ceilings on the revenues that are available. In addition, the tech team needs only support one version of the application, and the support team also only has one version to support. The vendor can push out updates, and all users will instantly be on the latest (and hopefully greatest) version of the application. In theory, at least, SaaS is a clear choice in almost every instance. However… SaaS is hard!

While the benefits to the business are many, the challenges that the team tasked with building the application will face are also many. And from the technologist's point of view, this is where things get interesting!

But before we dive into the details of SaaS applications, let's first consider what the alternatives are!

What other types of applications are there?

This book will describe SaaS, but while discussing what SaaS is and why it is a good choice in many instances, it will be contrasted with other types of delivery mechanisms for other types of applications. Some of the other *traditional* delivery mechanisms are discussed in the following sections.

Desktop application

This is the traditional type of application that was the major paradigm for many years. The software is packaged up into an *installer* and is somehow distributed to the end user. The distribution mechanism could be on a floppy disk or a CD or downloaded directly from the internet.

The application stores all of its files and data on the user's machine.

Typically, these types of applications are supplied with a product key that *activates* the software. Alternatively, licensing servers could be installed, allowing companies to more easily license the product on multiple computers.

On-premises web application

This type of application has largely been completely replaced with SaaS systems, but it was common for a while to develop a web application that could then be sold to multiple customers and installed on a server on the premises of the customer's organization.

This provided some of the benefits of SaaS but came with a lot of the baggage of a desktop delivery model.

The main benefit of an on-premises web application is that the purchasing company retains full control of the application. This means that they can choose to install an update—or not—based on an internal assessment of the costs and benefits of an updated version.

Another big plus of this delivery mechanism is that any data either stored in the database or transferred between the clients (web browsers) and the server can be retained entirely within the company's internal network infrastructure and need not ever touch the wider public internet. In theory, this does allow for additional data security.

A counterpoint to the aforementioned points: while it may seem more secure for a company to keep its data inside its own network, cloud services such as **Amazon Web Services** (**AWS**) from Amazon or Azure from Microsoft have put an incredible amount of resources into ensuring that stored data and data in motion is secure—indeed, their businesses depend on it. It is not guaranteed that a company's internal network is more secure.

While the ability to pick and choose versions and deploy updates at a time of the purchasing company's choosing may seem beneficial, this requires that the buyer employ people to install and manage the application, as well as to support their users.

What is "the cloud"?

Core to SaaS and delivering applications in the browser is the concept of *the cloud*. At its most basic, the cloud is just someone else's computer… but while accurate, that doesn't really quite do it justice. The cloud enables applications to run with very high levels of availability, and with essentially infinite—and instant—scalability.

For a SaaS app, the cloud is absolutely crucial. In almost every case, every component (database, **application programming interface** (**API**), and **user interface** (**UI**)) will be hosted using a cloud provider. The most commonly used cloud providers are Microsoft Azure, Google Cloud, and Amazon AWS.

The large cloud providers are essentially just providing access to an enormous amount of computing power, for a cost. This is termed **infrastructure as a service** (**IaaS**) and is a very important pillar of SaaS.

Where did it all begin?

While this is not a history of programming book, I think that it's worth taking a short look at the last 60-ish years of application development to see how we have arrived at the place that we have.

My first introduction to professional programming was visiting my father's office sometime in the late 1980s when I was around 6. I was shown the computer room, which contained an enormous VAX "mainframe" machine that did all the thinking. I was then shown around the offices where workers had terminals on their desks (complete with green text, just like in *The Matrix*) that connected back to the mainframe. The cost of the actual computer was such that there could be only one, and everyone simply got dumb terminals on their desks that looked something like this:

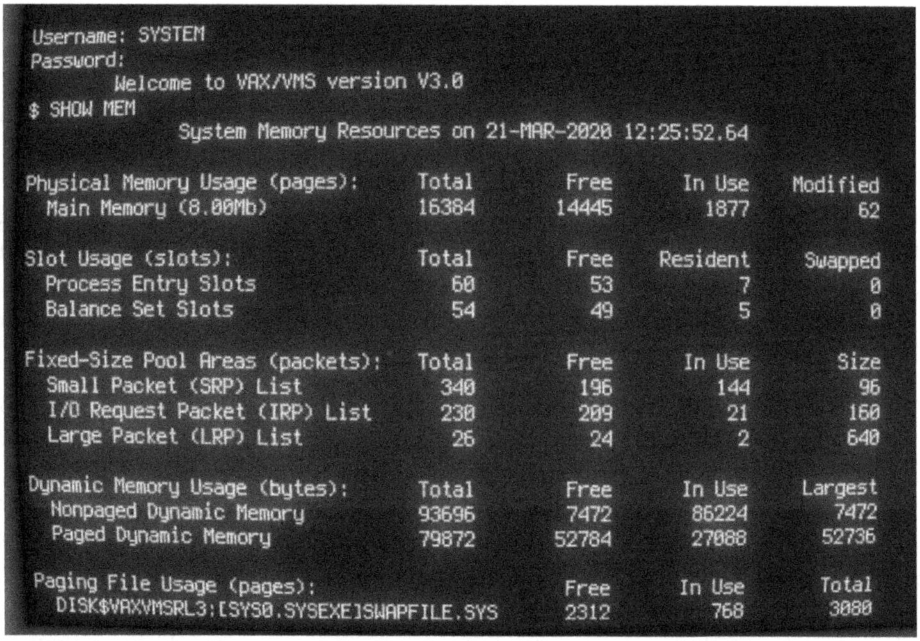

```
Username: SYSTEM
Password:
        Welcome to VAX/VMS version V3.0
$ SHOW MEM
        System Memory Resources on 21-MAR-2020 12:25:52.64

Physical Memory Usage (pages):    Total      Free      In Use    Modified
  Main Memory (8.00Mb)            16384      14445       1877          62

Slot Usage (slots):               Total      Free     Resident    Swapped
  Process Entry Slots                60        53          7           0
  Balance Set Slots                 54        49          5           0

Fixed-Size Pool Areas (packets):  Total      Free      In Use      Size
  Small Packet (SRP) List           340       196        144          96
  I/O Request Packet (IRP) List     230       209         21         160
  Large Packet (LRP) List            26        24          2         640

Dynamic Memory Usage (bytes):     Total      Free      In Use     Largest
  Nonpaged Dynamic Memory         93696      7472       86224        7472
  Paged Dynamic Memory            79872     52784       27088       52736

Paging File Usage (pages):                   Free      In Use      Total
  DISK$VAXVMSRL3:[SYS0.SYSEXE]SWAPFILE.SYS   2312        768        3080
```

Figure 1.1 – A VAX terminal

Despite the simplicity of the preceding display, I was instantly captivated, and this experience clearly resonated with me as I followed in my father's footsteps and became an application developer myself, albeit working with very different tools!

We were a long way off SaaS in the 1980s (although the "mainframe" and "terminal" configuration was perhaps instructive in some of the thinking that came later)!

While internet technology has existed since the 1960s, it has not done so in a way that is recognizable to the current generation of TikTok-addicted teenagers. The internet was little more than a technical curiosity until deep into the 90s and was far from the bedrock for delivering applications that it now is.

A huge technical pillar required to build a SaaS application was achieved in August 1994, when Daniel Kohn made the first secure credit card transaction. He bought a Sting CD—there is no accounting for taste! In November of the same year, Netscape Navigator introduced the **Secure Sockets Layer (SSL)** protocol, making it technically possible for anyone to transact over the internet without fear of having their information stolen.

Almost immediately building upon this new payment tech, Amazon launched the following year and was joined shortly by eBay. This was the very beginning of people willingly typing their credit card details into the internet with some degree of trust, but this practice was so far from mainstream at this point. (Sidenote—I made my first online purchase on Amazon in June 2001, buying *Happy Gilmore* on DVD; there is no accounting for taste!)

The first true SaaS app did not take long to emerge. Salesforce released what is considered to be the first true SaaS app, with its CRM platform launching in 1999.

However, to say that SaaS *started* in 1999 is not really giving the full picture. Sure—Salesforce was miles ahead of the curve, and its investment in its SaaS app has surely done wonders for the company (and its balance sheet) over the following decades. But the internet was still a curiosity to the vast majority of the population in 1999. While Salesforce could lean on its corporate clients with (relatively) fast internet connections, the reach of a SaaS app was tiny. Relatively few households had the internet, and none had broadband connections. The huge majority of the population would never consider putting their real name on the internet, never mind their credit card details! The era of SaaS ubiquity was still some time off.

By the time I started my career as a developer in the early 2000s and entered a corporate programming environment for the first time since the 80s, things had moved on considerably. There were no more VAX mainframes. By now, we all had blazing fast (by comparison) computers on our actual desks, so there was no need to delegate the computations to a centralized mainframe computer anymore. You could do the computations right there on your own computer. And so, we did! Throughout the 90s, thousands of developers churned out millions of "gray box" desktop applications, often written in VB6 with arcane installation guides that (hopefully) covered any and all eventualities. This was not really a high point for application development, certainly not enterprise applications, which constituted the vast majority at the time.

Round about this time, the internet was starting to mature to the point that it was a day-to-day business tool and was rapidly becoming as common as running water in the typical household. But even in the mid-2000s, the concept of "web apps" or Web 2.0 was still to emerge into the mainstream. Sure—there was PHP and ASP, and you could use those technologies to build web-delivered applications. But these were often more akin to clever websites, rather than what we would currently consider a fully-fledged "web application" these days. Despite the fact that online payments were becoming very common, there was still not really the concept of paying a monthly fee for a "service." The expectation was still that you would "buy" and therefore "own" and subsequently "install" software yourself.

This would, of course, change over the following two decades.

ASP.NET took what had been started with "classic" ASP and really ran with it. WYSIWYG editors were provided with Visual Studio, making the workflows very familiar for anyone who had been building the now old-fashioned "gray boxes" with VB6. "The Enterprise" almost immediately embraced "web apps" to replace the gray-box apps that had become typical and often universally hated.

This move from the enterprise to deliver software in the browser taught a generation of office workers about the benefit of web-delivered software, and before too long, they would start demanding that for all applications, even ones used in their personal life.

Email was most likely the first service catering to an individual that went fully web-based. The Gmail beta launched in 2004 with a then huge storage capacity of 1 GB for free… with more available for a cost! The monthly "subscription" for consumer-grade software was born.

These days the consumer market for SaaS applications is huge, with a multitude of task-list, note-taking, journaling, and email services provided. Not to mention that entertainment is now supplied "as a service" with monthly subscriptions to Netflix, Spotify, and many others now in many—if not most—households.

There are no residual concerns about entering payment information on the internet. There are barely any homes not serviced by broadband internet, and even developing countries often have robust mobile data networks and smartphones in the hands of many. There are no longer any barriers in place to delivering cloud-based applications over the web, for a monthly subscription.

It is clearly apparent that SaaS has eaten the world and taken over as the dominant method of delivering software! The era of SaaS is upon us!

Why is SaaS a popular choice for businesses large and small?

SaaS is becoming an increasingly popular choice across many different industries and many different sizes of enterprises. It is becoming ubiquitous across all manner of business models. There are a number of reasons for this, but for the most part, this is derived from the ability to add value to the user and revenue to the vendor. It is a win-win paradigm.

All manner of different applications can be developed and delivered over the internet and in the browser. The same tech stack—and, therefore, the same development team—can be used to deliver essentially any kind of application that the business can dream up. This makes it almost equally appealing to a start-up as it does a multi-national.

Using a traditional desktop application model or on-prem web app, the business needs to have a direct line to all of its customers for invoicing, providing updates, and so on. Acquiring new customers requires a sales team, and deploying new instances for new users requires a tech team. Given that each install is in a different environment, it is very likely that a sizable support team is also required to help with a myriad of different installations.

All of this melts away with SaaS. The only scaling consideration is the availability of virtual servers, which are near infinite in the era of Azure and AWS.

Because SaaS adds value for both the user and the vendor, it has become an obvious choice for businesses of all sizes.

Which tools are required to build SaaS apps?

The range of tools that could be used to develop a SaaS application is huge. The nature of developing a SaaS application is such that specific tools will be needed to build and test databases, APIs, and frontends, as well as many auxiliary tools such as static analysis, build pipelines, source control, and containerization tools, to name but a few.

This book will focus on the Microsoft tech stack, and as such will primarily use tools from Microsoft. But we will use a lot of them—such is the nature of building a SaaS app.

Database development

Starting at the bottom of the tech stack, we will use SQL Server Express for all database work. This can either be installed directly onto your developer machine or run in a container using Docker. Both methods will be described in detail in this book, although we will generally prefer containerized solutions.

API development

An API is a set of rules and protocols that specifies how two systems should communicate with each other. It is a way for one system to access the functionality of another system, such as a web-based software application or a server. APIs allow different software systems to interact with each other and share data and functionality. They are often used to enable integration between different systems, such as a mobile app and a backend server, or a website and a database. The API will be developed using C# and .NET 7. Sticking with all things Microsoft, we will use Visual Studio Community Edition. This is a free and extremely powerful IDE that makes developing the C# application very straightforward.

Frontend development

There are a plethora of good options with which to build the frontend for a SaaS application. The selection will largely come down to personal choice or the availability of developers on the market.

At the time of writing, by far the most commonly used frontend technologies are JavaScript-based—typically, Angular or React. However, the future is WebAssembly, and Microsoft has recently released Blazor using this technology, which allows the frontend to be built using familiar .NET languages, such as C#. This book will demonstrate a frontend using Blazor, but given that the JavaScript frameworks are (for now) more popular, I will take care to keep the explanations generic enough that the learnings with Blazor can be applied to any other frontend framework.

Authentication and authorization

Regardless of which frontend technology is used, it is of vital importance to get the authentication and authorization correct. We have dedicated an entire chapter to this later in this book. We will use an implementation of the OAuth 2.0 protocol, and will demonstrate how to secure your SaaS application from the UI all the way through to the database, and back again!

Hosting

Every SaaS application needs somewhere to live, and that will typically be in the cloud. While the bulk of this book will focus on developing on a local developer machine, we will also investigate how to build a deployment pipeline, and show your app "going live." We will use the Azure portal for all application and database hosting.

Docker

Developing a SaaS application really is the epitome of "full stack." We will be working with a wide array of tools, from databases to the frontend, not to mention many different testing frameworks to test all of these components. Accordingly, we will lean on Docker a lot to wrap up all of these dependencies and simplify the development process. Docker is a truly huge topic on its own, and it is outside the scope of this book to fully explain what Docker is and what Docker does. Put simply, Docker allows all sorts of complexity to be wrapped up in a very simple-to-use container.

For example, consider executing a few thousand unit tests against a UI and an API, and maybe a few integration and **end-to-end** (**E2E**) tests as well. There can be many dependencies involved in running these tests locally, and it can often take some time to configure a developer machine to successfully execute the test suite.

With Docker, it is possible to encapsulate a full testing suite within a Docker container to run the tests with a very simple Docker command. Further, these tests will run identically on any machine with the Docker client running. So, the Dockerized test suite will run just as happily on a Mac, Windows, or Linux, as well as in a pipeline on a cloud server.

In a nutshell, Docker wraps up complexity and facilitates simple interactions with complex systems.

Which techniques will be used to build SaaS apps?

Really, there will be no specific techniques that we will use that are not used in developing any other type of software application. However, I will briefly mention the techniques that I will use in this book.

Test-driven development (TDD)

One of the huge benefits of SaaS is that the application can be updated very quickly and essentially rolled out to each and every user at the click of a button.

This is great assuming that everything works as expected, but it is much less good if there is a bug in the code. Of course, we could build an extensive set of manual regression tests and build business processes into the release pipeline… but in doing so, you are losing a lot of the supposed advantages of SaaS—the ability to release often and early.

The only way to facilitate rapid deployments with some level of confidence that they will work is to build automated tests. And really, the best way to build an automated test suite is to do it as you go along, by following a TDD approach.

I am aware that TDD has a somewhat mixed reputation in the industry at present. In my opinion, that is because TDD done wrong is a nightmare, and TDD is done wrong very often. I will present a variety of TDD that I feel is an excellent support structure when developing SaaS apps.

Domain-driven design (DDD)

DDD is defined as an approach to software development where the problem is specified by domain experts and not by middle management.

DDD is a software development approach that concentrates on understanding and modeling the business domain of an application in order to improve the design and implementation of the software. It emphasizes the importance of domain knowledge in software development and encourages the use of domain-specific language in the design and implementation of software systems.

In DDD, the business domain is understood as the core area of expertise or focus of an organization, and the software being developed is viewed as a tool to support and enhance the work being done in that domain. The goal of DDD is to create software that is aligned with the business needs and goals of an organization and that accurately reflects the complexity and nuance of the business domain.

SaaS products are often simply made available to anyone with a web connection, and there is no dedicated sales team individually approaching every customer and actively selling the product. Therefore, the product must sell itself, and so it must be useful. In order for this to be true, it is essential that the product is meeting a specific user need and addressing a particular problem domain.

Microservices

SaaS projects need to be flexible so that the product can evolve with the market and with customers' demands. It is very important that the product is architected in such a way that allows the straightforward addition of new features, with a minimal impact on existing features. A microservice-based architecture fits this requirement.

Multi-tenancy

Because every user is a tenant in the same deployed version of the application, the users' data must be kept separate in the data storage and retrieval systems. There are a number of ways to approach this, which are discussed in a subsequent chapter of this book.

Reactive design

A SaaS application lives online and is accessed through a browser. In the modern era of smartphones and tablets, there is no way to know what type of device will be used to access your application. The frontend really needs to work on any type of device, or at the very least "fail gracefully" if it cannot operate on any given device.

Accordingly, the design of the UI must be "reactive," meaning that it can be rendered in a way that is befitting the device that it is being displayed on.

Progressive web apps (PWAs)

Something "nice to have," but I feel it is worth considering. When we are building SaaS apps, we really want the user to feel that they are using a full-blown "application" and not a glorified website. However, by definition, a website cannot be shown if there is no available internet... PWA-based designs work around this by allowing limited functionality to work where there is little or no internet available.

Of course, with no access to the backend, many of the functions of the site will be unavailable. There is no way around that, but PWAs can be used to make that a little less painful for the user, and so it is an important technique that authors of SaaS applications should be aware of.

We will demonstrate a PWA using Blazor for the frontend technology.

Reasons for choosing SaaS as the go-to paradigm for all manner of businesses, large and small, old and new: as we have highlighted previously, much—if not most—of the population is using some form of SaaS application, from Gmail to Netflix. If SaaS is really eating the world, there must be a reason for that.

No installations needed

This is the biggest and most important thing that drives businesses toward a SaaS-based solution.

Using a traditional application model, a new customer will often have to contact a sales team and sometimes also a tech support team to get the software installed on-premises and appropriately licensed.

With SaaS, a new customer can discover the app and sign up for an account in seconds without needing any contact from the company that is providing the application. This is a significant saving for the company that has developed the application in that no additional sales or support teams are required to onboard the new customer. This also prevents a time lag from discovery to installation, during which the customer could change their mind or discover a competitor.

Browser-based delivery

The users of a SaaS application are not limited to accessing the application on a specific computer or in a specific network environment where the license server resides. The users can have access to the application from any internet-connected machine anywhere in the world. In the modern era of smartphones and tablets, the user of a SaaS application may not even need a computer to take full advantage of the provided service.

Scalability

In a traditional model, this is simply not possible. Convincing people to install any application on their computer is hard. Convincing companies to do so is doubly hard. Scaling up the user base of a desktop application requires a dedicated sales team and also a dedicated support team to first make the application, and then hand-hold people through the installation process. With a SaaS app, users simply sign up on an online form.

Because the SaaS application is hosted in the cloud (Azure, **Google Cloud Platform** (**GCP**), or AWS), near-instant scale-up of the infrastructure is possible should a sudden spike in demand occur. There is no other software delivery paradigm that could face an overnight 10 times increase in demand and not leave the vendor floundering!

Upgradability

Under the "legacy" methods of delivering applications, upgrading the apps tended to become nightmarishly complex and expensive for the vendors of the apps.

If you consider a traditional desktop application, the same application could be installed on hundreds of machines across hundreds of different businesses. No two businesses will be running the same hardware or OS versions, and so it is impossible to keep everyone on the same version of your software. It is similarly completely impossible to roll out an upgrade to a new version. At best, you can withdraw support for a particular version at a particular time and hope that everyone stops using it (note—they will not stop using it).

This is a similar problem for an on-premises web application. While there are fewer induvial installations, there will still be many different versions out there, servicing specific business requirements.

This problem completely melts away when you move to a SaaS paradigm. The vendor has full control of the upgrades and can roll them out to all customers all at once at the click of a button.

Only having one version of an app to support is a *huge* saving for the vendor and also a huge advantage for the development team.

Iterate quickly

An emergent property of the upgradability advantage of a SaaS app is the ability to iterate on the solution extremely rapidly. Feedback about new features can be incorporated and pushed out to all users in very quick cycles. This allows for a very fast turnaround from a developer writing a feature to that feature providing value to the users and revenue to the vendor.

Consider how long it takes for that value to be realized in a traditional app. The code may well sit in a Git repo for many months before it is even included in the "annual release," and then, the users may choose not to upgrade immediately.

Analytics

Because the users are all funneled through a single web application, it is extremely easy to analyze *how* the application is being used. This can guide the business to make smart decisions to upgrade the most used parts of the application and defer working on less well-used parts. Couple this with the ease of upgrading and the fast iterations, and this can provide a huge boost in value to the users and should come with a boost in revenue to the vendor.

Global market

With SaaS, the ability to reach customers anywhere is a massive boost for the business supplying the product. There is no case where a sale is missed due to time-zone issues or a sales rep failing to reply to an email.

The ability to gain access to a truly global audience has allowed some companies to become the biggest companies in the world—bigger than banks and oil and gas majors. This access to a market also allows thousands of smaller companies to thrive in a truly global environment.

Flexible payment model

There are many different payment models available to a business offering SaaS. This allows it to capture customers large and small and derive the most value possible from each tier or size of the customer. Some types of payment models follow:

- Tiered pricing
- Freemium
- Free trial periods
- Per-user pricing

Security

Access to the SaaS application will be through a user login that protects sensitive data and files, which are securely stored in a cloud system. This is (in most cases) far more secure than storing data on a server on-premises—or on a local user machine, in the case of a desktop application. While it may seem that keeping all of the data locally or within a local network is more secure than sending the data to a server on the cloud over the internet, this is often not the case. A huge amount of effort goes into securing the cloud services that are typically used to host SaaS applications, and often that effort cannot be replicated on a site-by-site basis, whether that be securing an on-premises web app or individually securing the data on desktop-installed applications.

How does this affect the development process?

In theory, the development process for building a SaaS app is rather similar to any other type of application, but in practice, there are nuances and considerations that you must note as a developer working on a SaaS project.

Release often, release early

There is no "annual release cycle" for a SaaS app. The expectation is that functionality will be broken down into manageable slices, developed, and then pushed out as soon as they are ready. This requires a bit of a mindset shift if coming from a more traditional release cycle. All changes have to be incremental, generally contained to a small part of the applications, and ready for release ASAP.

This mindset of getting the code into the hands of the users will have to be backed up by automated pipelines that build and release the new code without much manual intervention.

While it is possible to roll out the updates to sections of the audience to make sure there is nothing catastrophic in there, it is much more typical for a small-to-medium-sized SaaS application to simply push the updates out to all of the users all at once. For this to be successful…

Testing, testing, testing

If the code is released to the entire user base in one go, and often with no real way of rolling back the change, you had better hope that it works!

The only way to build any sort of trust that the code will operate as advertised is to test it, and in a "release often, release early" mindset, this means automated testing. While this doesn't necessarily mean adhering to a TDD mentality, this can certainly be useful.

You'd better be full stack

Okay—this is not absolutely required. I'm sure that the larger SaaS applications are worked on by specialists in the database/backend/frontend disciplines. But it will certainly help to have a good knowledge across the different application layers.

The way that SaaS applications "grow" in an almost organic way through the fast cycles and near-instant releases means an understanding at least across the layers, and an understanding of how a decision in the database *may* affect the frontend is invaluable.

Know thy user

While this may not be required for each developer in the team, it is absolutely essential that the team as a whole understands who is using its product, why they are using it, and where the value is. This knowledge and understanding will come from assessing the analytics and also from "softer" approaches, such as focus groups and interviews with users.

This understanding should flow down into the development process through user stories. From the developers' point of view, this may manifest in seemingly sudden shifts in direction, if a particular feature has not landed well with a focus group or user interviews have shown that a particular path is the wrong one. The ability to pivot quickly is important across the team.

What to expect from this book

This chapter has served as an introduction to what SaaS is, where it has come from, why businesses and users love it, and finally what is required of you as a developer to effectively build a SaaS application.

In the coming chapters, we will deep dive into all of the aforementioned areas, but with a focus on building the tools and understanding required as developers to build great SaaS applications that your users will love to use, and that (hopefully) you will love to build!

To illustrate the technical points and understanding required, we will build, test, and deploy a full stack SaaS application!

I have approached the writing of this book with the same mindset as I do when I am building an application. I aim to make this an engaging, interesting, and maybe even "delightful" experience for the user—which is you in this case!

Let's get going!

How to assess and pay off technical debt

What is technical debt anyway? Let's start with a definition:

Technical debt accumulates as features are added to a project. It is inevitable that as complexity is added to certain areas of a project, some of the other parts of that project will no longer fit quite as well and will at some point have to be worked on. However, the realities of life as a developer are that products must be shipped and that bills must be paid, and therefore the time is not always allocated to tidy up every last corner, and over time, technical debt starts to accumulate.

Another source of technical debt that all projects will experience occurs when the underlying frameworks and technologies are updated. It is often not straightforward to update a major release, and sadly it is common for projects to languish in outdated versions—which represents a technical debt.

One final common source of technical debt is when users' habits change. In the era before the iPhone, very few people accessed a website through a mobile device. That very quickly changed, leaving many teams scrambling to update their websites so that they functioned correctly when accessed through a mobile device.

All technical projects have some technical debt—there is no getting away from that fact, and it is important for all projects to keep on top of this. However, there are a few considerations specific to developing SaaS applications that must be taken into account.

The philosophy with most SaaS applications is to get the development work into the hands of the customers as quickly as possible. This is usually achieved through extensive automated test suites, coupled with build and release pipelines to push the code out to the production environments as soon as possible.

Contrast this with traditional release and delivery mechanisms, where there will be a year gap between releases, and (hopefully) some time in that year allocated to paying off technical debt.

With the constant develop-release-repeat cycle that is common to SaaS development, it is important that technical debt is kept on top of.

The first and most important way to assess and pay off technical debt is to allow the development team some time each week (or each sprint cycle) to do "chores." These would be small housekeeping tasks that might otherwise be left to grow into serious technical debt issues. The development team are *always* the people who know best where the technical debt is. After all, they created it in the first place!

Static analysis is another extremely powerful tool to keep on top of technical debt. Static analysis is used to check the quality of the code when it is not running (when it is static!) and can check that standards are adhered to and the latest best practices are being implemented.

Similar to static analysis, linting should also always be performed to ensure that the code is formatted according to the agreed coding standards.

As mentioned previously, out-of-date packages can become a major source of technical debt. While being on the absolute cutting edge can be risky, there is rarely a benefit in being significantly out of date. There should be regular housekeeping done to ensure that any packages and frameworks that are being used are sufficiently up to date.

Finally, automatic performance testing should be carried out to ensure that there are no regressions in the performance of the application as it grows and changes over time.

Even if all of the aforementioned points are strictly adhered to, a project will still accumulate technical debt over time. There is very little that can be done about this. But with the preceding considerations and mitigations in place, the impact that technical debt has on the project—and, ultimately, the profitability of the company building the project—can be minimized.

Summary

This chapter has given a broad introduction to the concept of SaaS. We have covered a brief history of application development and looked at the paradigms that were prevalent in the pre-SaaS days. We have considered why SaaS is becoming so popular and looked at the technical, business, and user-centric reasons for its adoption. Finally, we have considered which tools and techniques you as a developer need to be effective as a SaaS developer.

Hopefully, this chapter has given you a strong foundational understanding of SaaS, which we will now start to expand on as we move through this book!

Building SaaS applications is challenging, but by far the best way to make progress is to start building and get your hands dirty! In the next chapter, we will get straight into the tools, tooling, and techniques that will be used throughout this book to build a demo of a SaaS application!

Further reading

- For current business trends: `https://www.datapine.com/blog/saas-trends/`

- For some SaaS development tricks and tips: `https://www.rswebsols.com/tutorials/software-tutorials/saas-application-development-tips-tricks`

- For further information on scaling SaaS applications: `https://medium.com/@mikesparr/things-i-wish-i-knew-when-starting-software-programming-3508aef0b257`

2

Building a Simple Demo Application

In this chapter, we will build a very simple demo application that will form the basis and framework of the SaaS application that we will build out in subsequent chapters. The purpose of this chapter is to get familiar with all of the tools and techniques that we will leverage later in this book, and in our careers as builders of SaaS applications! Once we have this foundational knowledge and framework app, it will be much easier to start to build more complex functionality.

Before we start building the framework application, we must first discover and install the tools that we will use. Then, we'll build out the skeleton application. Once we have that working, then we will start to think about what would be required to flesh this out into a real-world SaaS application!

This chapter will give you a very brief look at a lot of different technologies. Please don't feel overwhelmed! We are going to configure a dev environment from scratch, using Docker, and then we are going to initialize a database, API, and UI.

This chapter uses a lot of different technologies, some of which you may not be familiar with. Don't worry! Everything in this chapter is written out in detail, and anything that is skipped over quickly in this chapter will be explained fully in subsequent chapters.

Developing SaaS applications, almost by definition, requires a wide array of different technologies. We will take a quick look at them all in this chapter, and then deepen our knowledge and understanding in subsequent chapters!

This chapter covers the following main topics:

- Getting set up – installing the tools
- Building a simple example application
- Where do we go from here?

By the end of the chapter, you'll have built, installed, and configured your development environment, and you'll have initialized all of the components required to build out a SaaS app. Hopefully, you will finish this chapter inspired to understand how we will make the most out of these tools and techniques!

Technical requirements

All code from this chapter can be found at `https://github.com/PacktPublishing/Building-Modern-SaaS-Applications-with-C-and-.NET/tree/main/Chapter-2`.

Setting up

In this section, we'll cover the tools that I recommend using to work through the examples in this book. Please note that there are not really any specific tools for building a SaaS application – the typical set of developer tools that you are probably accustomed to will do the job. However, I will refer to a specific set of tools in this book, and so I'll describe them in this section, explain my selection, and cover how to install and configure the tools.

I am sure that any of the popular tools out there would be perfectly sufficient, so don't feel like you *must* go with my selection, and feel free to use whatever you are most comfortable with! You should be able to adapt the examples to work with your tooling of choice.

There are only two tools that need to be installed on your machine to follow along with the code that will be supplied in this book:

- Docker Desktop
- Visual Studio Code

These tools are fairly stable, and it is likely the best choice to use the latest version. For reference, I am using Docker v20.10.17 and VS Code v1.71.2.

These tools are chosen deliberately as they are available for all operating systems, they are free to use, and they will give us a consistent base on which to work through the code examples. These are the two tools that I find most useful in my day-to-day work as a software developer! They are typically the first two applications that I install when building a new developer machine.

I'm going to use a neat trick with Docker to containerize the entire developer setup, which is why these two tools are the only ones that you need to install on your machine, and it is also the reason that I am confident that the example provided will work on any operating system.

Visual Studio Code

Visual Studio Code has gone from being a *lite* version of the full Visual Studio application to being an absolute workhorse for all manner of development work. With the availability of a massive range of extensions, it can be configured for just about any coding task, and then further modified to suit individual preferences.

VSCode has become my IDE of choice for many different projects, but the real strength of a SaaS project is that it can be configured to support all of the different layers that will be built. You can add extensions for database development, API development, and UI development and build a highly customized developer environment that is specifically tailored to your project and your preferred tooling.

Start by installing Visual Studio Code. This is done simply by clicking this URL and following the instructions for your operating system of choice: `https://code.visualstudio.com/download`

When VSCode is installed, you will also need to install a couple of extensions.

You will need three extensions for VS Code:

- Remote containers

- Remote WSL

- Docker

The tags for the required extensions are as follows:

```
ms-vscode-remote.remote-containers
ms-vscode-remote.remote-wsl
ms-azuretools.vscode-docker
```

When VSCode is correctly configured, the **EXTENSIONS** pane should look like this:

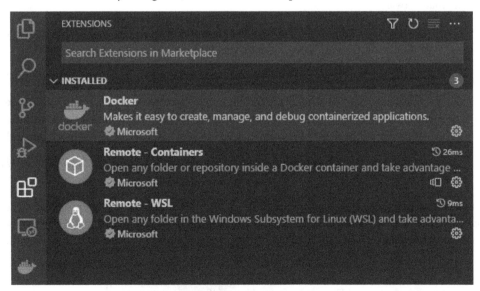

Figure 2.1 – Visual Studio Code with extensions configured

We will use a number of other extensions specific to the project, but these will be installed in a dev container for ease of use – more on this later!

Docker

Docker is a relatively recent addition to the developer toolbox, but it has quickly become completely invaluable for building and deploying applications. In more recent years, Docker has also become an incredible tool to encapsulate all of the setup required on a developer machine.

Docker is a containerization tool. The principle at play is derived from shipping containers. Shipping containers are a standard size, which means that the machinery to move them and load them at the port and onto ships is completely standardized around the globe. Of course, what goes into the shipping containers can be anything from cars to cucumbers, but the infrastructure to move the containers is identical regardless.

This principle is applied to software packaged with Docker. Whatever software tool is to be run is put inside a Docker container – analogous to a shipping container – and the Docker infrastructure is then used to run the container.

You can consider a container to be a tiny virtual machine, or maybe an executable of its own. The important thing to keep in mind is that any piece of code, application, service, and so on that has been placed inside a Docker container will run on any computer with Docker installed. Furthermore, it will run (or not run!) in an identical way regardless of whether or not the host operating system is Windows, Mac, or Linux.

This concept of containers is fantastic for shipping running software. I can now pull a Docker image for just about any software utility I can think of, and no matter what platform I am working with, I can execute that software.

From a software provider's point of view, the huge benefit is that they can be certain that their code will run correctly on my machine because it's running against a standard platform – Docker!

You may be wondering why this is important when it comes to setting up a developer environment.

Dev containers are a relatively new but extremely powerful concept that can take all of the power that Docker has when running applications and apply that to developing applications. It is now possible to configure an entire dev environment – with all of the associated dependencies – and put all of that in a container!

This may not be of the utmost importance for a more basic project, but a SaaS application is rarely a basic project.

The nature of a SaaS application typically requires that many project-specific tools are installed on each developer machine. Worse, it is very common that specific versions of the tools will be required, which can often make the initial setup for new team members particularly painful. These tools can include the following:

- A database platform
- A framework, such as .NET – very often with specific version requirements
- Package management, such as NPM or NuGet
- A web server of some kind for the frontend

- A huge number of CLI tools
- And many other dev tools

If your dev team supports multiple projects – as is often the case – this can become a real pain point.

I try to use dev containers as much as I possibly can to wrap up dependencies, and I will use this technique in the examples in this book.

Please note that Docker is an enormous topic in itself, and it is out of the scope of this book to cover it in any sort of depth. I will cover just enough to get familiar with the techniques that I am using, and leave it up to the reader to do more of a deep dive into all things Docker!

Dev containers

As developers, we all have our favorite set of tools that we work with on a day-to-day basis. This will start with our choice of operating system, and from there we will pick IDEs, database platforms, frontend frameworks, and all of the associated tooling.

This huge variety of systems, tools, platforms, and frameworks provides a challenge for the author of a book describing how to build a SaaS platform…

In order for the tutorials in this book to appeal to the broadest range of developers possible, I am going to make use of a relatively new concept known as **Dev Containers**. This allows us to configure a Docker container to do all of our development. This will give you a consistent platform with which to work and will ensure that all of the examples that are provided will work on literally any machine.

Making use of dev containers gives us a completely consistent development platform to work with so that we can be sure that all readers of this book will be able to run the example code that is provided, but there is a small configuration overhead to get started.

What is a dev container?

Continuing with the analogy of a shipping container, a dev container is simply a wrapper around the myriad of developer tools that you will use when working on the code in this book.

Broadly speaking, the tools and services that will be required are the following:

- A database platform
- The .NET SDK
- Extensions to support database development
- Extensions to support API development
- Extensions to support frontend and Blazor development

All of the above will be packaged up in a dev container.

Configuring the Docker containers

The code for this first example is available at `https://github.com/PacktPublishing/` `Building-Modern-SaaS-Applications-with-C-and-.NET/tree/main/Chapter-2`. Either clone the repo or follow through with this example setup.

If you are following along, then start by creating a new folder that will form the root of your project, and then open that folder in VSCode. This is a completely blank canvas, but by the end of this chapter, we will have the skeleton framework for a functioning SaaS application.

Start by creating the following folder and file structure:

Figure 2.2 – Expected folder structure

With the folder structure created, we can now start to fill out these files. We'll start with the Dockerfile in the `dev-env` folder. This will be used to configure the developer environment and will contain the instructions to install the tools that will be used to build the application.

Dockerfile

Open the file:

```
docker/dev-env/Dockerfile
```

And paste in the following:

```
# [Choice] .NET version: 7.0, 6.0, 5.0, 3.1, 6.0-bullseye,
5.0-bullseye, 3.1-bullseye, 6.0-focal, 5.0-focal, 3.1-focal
ARG VARIANT="7.0"
FROM mcr.microsoft.com/vscode/devcontainers/dotnet:0-${VARIANT}
# [Choice] Node.js version: none, lts/*, 16, 14, 12, 10
ARG NODE_VERSION="none"
RUN if [ "${NODE_VERSION}" != "none" ]; then su vscode -c "umask 0002
&& . /usr/local/share/nvm/nvm.sh && nvm install ${NODE_VERSION} 2>&1";
fi
# [Optional] Uncomment this section to install additional OS packages.
RUN apt-get update && \
    export DEBIAN_FRONTEND=noninteractive && \
```

```
        apt-get -qy full-upgrade && \
        apt-get install -qy curl && \
        apt-get -y install --no-install-recommends vim && \
        curl -sSL https://get.docker.com/ | sh

RUN dotnet tool install -g dotnet-ef
ENV PATH $PATH:/root/.dotnet/tools

# configure for https
RUN dotnet dev-certs https
```

This configures the developer environment to facilitate .NET application development. Let's go through this in detail. The first line determines the version of .NET that will be used:

```
ARG VARIANT="7.0"
```

We are using 7.0, which is the latest version at the time of writing.

Next, the version of Node is configured:

```
ARG NODE_VERSION="none"
```

No node version is installed. It is often useful to enable node or npm, but they are not needed at present.

The next command is used to install any packages or tools that you may want to use:

```
RUN apt-get update && export DEBIAN_FRONTEND=noninteractive \
&& apt-get -y install --no-install-recommends vim
```

This command updates the package manager and installs the Vim CLI tool, which we will use later.

The above has all been a "standard" configuration of the environment. Next, we will add some bespoke customizations that will allow us to make use of Entity Framework:

```
RUN dotnet tool install -g dotnet-ef
```

This command installs the .NET **Entity Framework** (**EF**) tools that we will use to interact with the database.

Finally, we will add the dot net tools to the `path` variable so that we are able to make use of them from the command line:

```
ENV PATH $PATH:/root/.dotnet/tools
```

Sqlserver.env

This file simply sets some environment variables for the SQL Server instance that we will spin up shortly. Copy in the following:

```
ACCEPT_EULA="Y"
SA_PASSWORD="Password1"
```

It should go without saying that passwords should never be used in any environment other than this example! Also, note that checking in passwords is extremely bad practice, and should not be done when working on a production application. We will remedy this when we talk about security in *Chapter 8*.

Docker-compose.yaml

This is where things get interesting. Docker Compose is a tool that lets us use multiple containers at the same time. It is a container orchestration tool!

Paste the following into the docker-compose file:

```
version: '3.4'
services:
  sql_server:
    container_name: sqlserver
    image: "mcr.microsoft.com/mssql/server:2022-latest"
    ports:
      - "9876:1433"
    volumes:
      - habit-db-volume:/var/lib/mssqlql/data/
    env_file:
      - sqlserver/sqlserver.env
  dev-env:
    container_name: dev-env
    build:
      context: ./dev-env
    volumes:
      - "..:/workspace"
    stdin_open: true # docker run -i
    tty: true # docker run -t
  volumes:
    habit-db-volume: null
```

The preceding commands for Docker Compose tie everything that we have done together. With this script, we can start up a set of Docker containers that allow us to build the API, and the UI, and also interact with a database – all without having to install anything directly on the host machine!

To recap: we have constructed a container called dev-env that is specific to our requirements. We have also provisioned a container with an instance of SQL Server 2022 running in it that will serve as our database platform.

Running the containers

That is all of the Docker-specific configurations. Now we will run the containers and start to interact.

We have built a Docker-based environment – now it's time to start it up. To do this, open VSCode, open a new terminal, and enter the following:

```
cd .\docker\
docker compose up
```

Your terminal should look like the following:

Figure 2.3 – Expected output

This can take a while the first time you run this command, but it is significantly faster in subsequent uses. This command will start up the services that are described in the docker compose file. Namely, a SQL Server instance, and the developer environment that we configured with .NET 6.0 and Entity Framework.

With the above, we have created a container with .NET and Entity Framework installed. You can convince yourself that we have actually achieved that by trying the following. Open a new terminal in VSCode and enter the following to exec into the dev-env container:

```
docker exec -it dev-env /bin/bash
```

The exec command allows interactions with an already running container through the terminal.

Running the preceding command will open an interactive terminal in the dev-env Docker container where we can check to see whether the .NET and EF tools are correctly installed by typing the following into the console that was opened by the preceding exec command:

```
dotnet --version
dotnet-ef --version
```

The preceding commands should return the following:

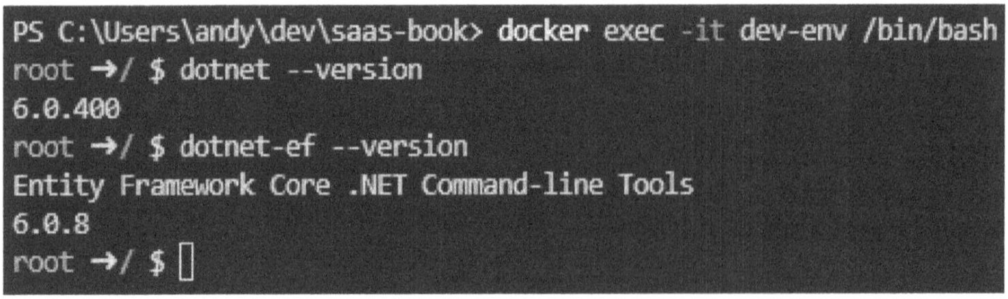

```
PS C:\Users\andy\dev\saas-book> docker exec -it dev-env /bin/bash
root →/ $ dotnet --version
6.0.400
root →/ $ dotnet-ef --version
Entity Framework Core .NET Command-line Tools
6.0.8
root →/ $ []
```

Figure 2.4 – Expected terminal output

Next, we can do something similar to ensure the SQL Server container is running. Again, open a new terminal in VSCode and enter the following:

```
docker exec -it sqlserver /bin/bash
```

Again, this will open another interactive terminal in the sqlserver container. You can convince yourself of this by typing the following:

```
/opt/mssql-tools/bin/sqlcmd -S localhost -U SA
```

And when prompted, enter the password from the sqlserver.env file (Password1) to enter the SQL Server command-line interface. You can do simple checks on the database platform, such as checking the version:

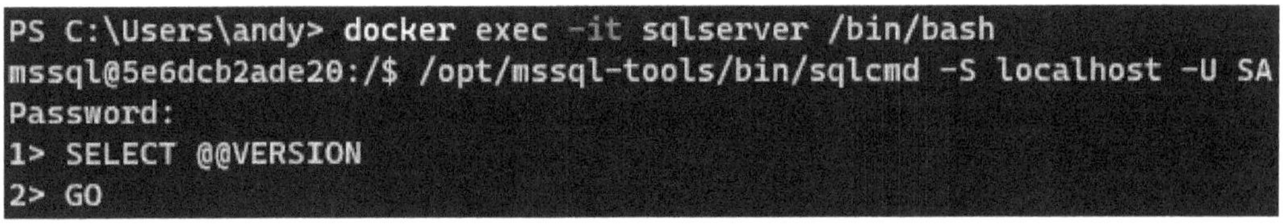

```
PS C:\Users\andy> docker exec -it sqlserver /bin/bash
mssql@5e6dcb2ade20:/$ /opt/mssql-tools/bin/sqlcmd -S localhost -U SA
Password:
1> SELECT @@VERSION
2> GO
```

Figure 2.5 – Checking the version

At this point, we have done all of the Docker container setups, but the only way that we can interact with our environments is to use the command line, and these days, that is not really acceptable! Thankfully, Visual Studio Code has a neat trick up its sleeve!

Configuring dev containers in VSCode

You will remember that we installed the Remote Containers extension for VSCode. This will allow us to open an instance of VSCode that makes use of the Docker containers that we have set up previously. This will require a little bit of additional configuration, but once set up, it will "just work" for the remainder of the project!

Start by creating a folder called `dev-env` in the project root:

```
.devcontainer
```

This is where VSCode will look to get the configuration.

In this folder, create a file called `devcontainer.json`. Your folder structure should look like this:

Figure 2.6 – Folder structure

In the `devcontiner.json` file, paste the following:

```
{
    "name": "SaaS Book",
    "dockerComposeFile": ["../docker/docker-compose.yaml"],
    "service": "dev-env",
    "workspaceFolder": "/workspace",
    "customizations": {
        "vscode": {
            "extensions": [
                "ms-dotnettools.csharp",
                "shardulm94.trailing-spaces",
                "mikestead.dotenv",
                "fernandoescolar.vscode-solution-explorer",
                "jmrog.vscode-nuget-package-manager",
                "patcx.vscode-nuget-gallery",
                "pkief.material-icon-theme",
                "ms-mssql.mssql",
                "humao.rest-client",
```

```
            "rangav.vscode-thunder-client",
            "formulahendry.dotnet-test-explorer",
            "kevin-chatham.aspnetcorerazor-html-css-
              class-completion",
            "syncfusioninc.blazor-vscode-extensions",
            "ms-dotnettools.vscode-dotnet-runtime",
            "ms-dotnettools.blazorwasm-companion"
            ]
        }
    },
    "remoteUser": "root"
}
```

Let's go through this line by line. The following line of code tells VSCode where to look for the `docker` compose file. This is the file that we created previously, which configures the two Docker containers:

```
"dockerComposeFile": ["../docker/docker-compose.yaml"],
```

The following line simply tells VSCode that the container named `dev-env` is the primary container:

```
"service": "dev-env",
```

The next line defines a working folder inside the container – more on this shortly:

```
"workspaceFolder": "/workspace",
```

The following is quite clever including the dev container configuration. This section allows us to define which extensions we want to have available to us when we start to work in the dev container. I have listed three extensions to get us started, but this list will grow as the project gains complexity:

```
"extensions": [ … ],
```

This is a particularly clever addition from Microsoft because this allows the VSCode configuration to exist as JSON within the project. This file is checked into the repo and moved with the code, meaning that any time a new team member pulls this repo, they will immediately have a fully configured editor that is specific to the project.

We have all become familiar with *infrastructure as code* in recent years. Using a dev container allows you to configure the *developer environment as code*. Doing this makes it incredibly easy to onboard new team members, and also means the end of *well, it works on my machine*. Using this technique means that everyone is working on a completely consistent platform, regardless of the physical hardware, the operating system choice, or the specific version of .NET or Node.

This is a huge win for developing SaaS applications, given the complex and often arduous route to getting a new team member up and running.

Starting the environment

That is the end of the configuration. I hope that we are all now working on a consistent and predictable platform, and will have no trouble following along with the examples in this book!

To start the environment in Docker, hit *F1* to open the command menu, and search for `Remote-Containers: Rebuild and Reopen in Container`.

You will see the options in *Figure 2.7* appear. Select the first one and hit *Enter* to continue.

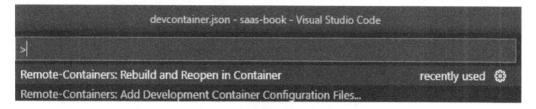

Figure 2.7 – Rebuild and Reopen in Container

This will close the current instance of VSCode, and reopen a new instance that is operating out of the `dev-env` container that was configured previously. Note that this can take some time the first time you do it, but it is faster in subsequent uses.

You will notice a few differences! First of all, look in the bottom-left corner, and you will see that you are running in a dev container called `SaaS Book`:

If you click to view the installed extensions, you will see a separate pane that shows the extensions that are installed in this instance of VSCode match the extensions specified in the `devcontainers.json` file. This list will start to grow significantly as the project takes shape throughout the course of this book.

Figure 2.8 shows some of the extensions that are installed in the container:

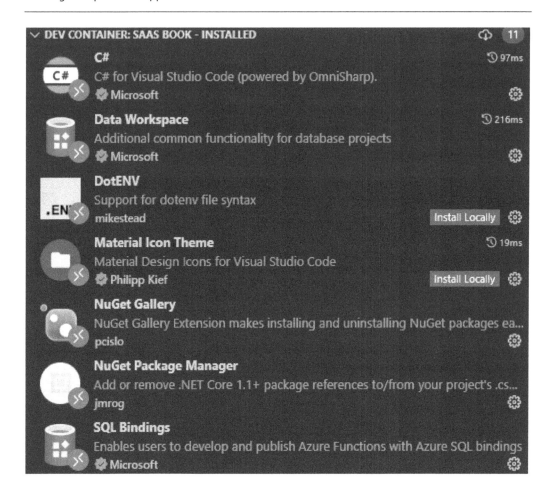

Figure 2.8 – Installed extensions

If you open a terminal, you will also notice that this is a Linux environment. You can convince yourself of that by checking the version by running the following code:

```
cat /etc/os-release
```

You'll see the following output:

```
root →/workspace (master X) $ cat /etc/os-release
NAME="Ubuntu"
VERSION="20.04.5 LTS (Focal Fossa)"
ID=ubuntu
ID_LIKE=debian
PRETTY_NAME="Ubuntu 20.04.5 LTS"
VERSION_ID="20.04"
HOME_URL="https://www.ubuntu.com/"
SUPPORT_URL="https://help.ubuntu.com/"
BUG_REPORT_URL="https://bugs.launchpad.net/ubuntu/"
PRIVACY_POLICY_URL="https://www.ubuntu.com/legal/terms-and-policies/privacy-policy"
VERSION_CODENAME=focal
UBUNTU_CODENAME=focal
root →/workspace (master X) $ □
```

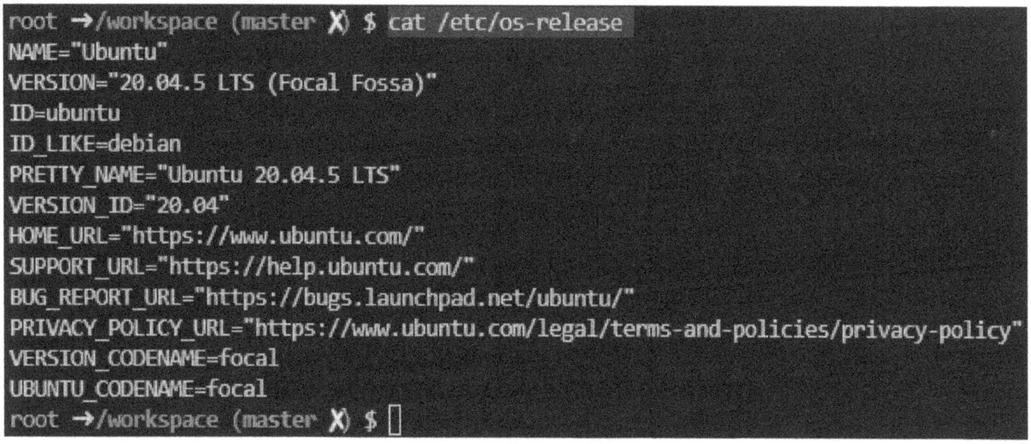

Figure 2.9 – Output

This shows that we are in a Debian environment. Notice also that we are working in a folder called workspace, which is what was configured in the devcontainer.json file previously. This is configurable as per your preferences.

To further convince ourselves that this is indeed the dev-env container that we configured earlier, you can check the .NET, and the Entity Framework versions again in the terminal in VSCode:

```
dotnet --version
dotnet-ef --version
```

You will see the following output:

```
root →/workspace (master X) $ dotnet --version
6.0.400
root →/workspace (master X) $ dotnet-ef --version
Entity Framework Core .NET Command-line Tools
6.0.8
root →/workspace (master X) $ □
```

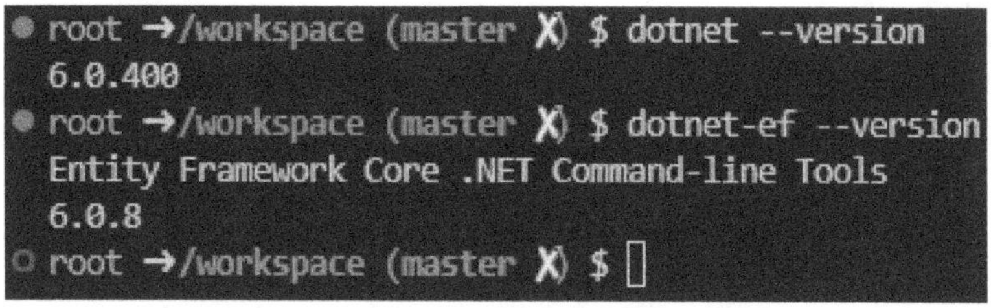

Figure 2.10 – Terminal output

That is the environment configured to the extent that is currently required. Let's recap:

- We have a container that has the .NET SDK installed, ready to use to build any kind of .NET applications.

- The same container has Entity Framework command-line tools installed so that we can use Entity Framework to build and interact with a database. You will recall that we have installed Entity Framework in the Dockerfile.

- We have a separate container running SQL Server, which hosts the database that we can interact with. It is worth noting that we have not installed SQL Server itself on the host machine. We have access to the database via a Docker container. You will see more of this in *Chapter 3* and *Chapter 4*.

This is now the basis of a developer machine that can be used to build SaaS applications, and we have achieved this without ever having to install any of these tools directly on the machine that you are working on – everything is packaged up in Docker containers. This configuration exists as code, and as such it moves with the repo. Any developer who clones this repository locally will immediately have access to all of the project-specific tools and configurations.

As we progress through this project, this setup will become more complex. It will grow as the project grows. But this is sufficient for now, so we will leave it here and start to piece together a very basic application.

Configuring VSCode

At this early stage, there is not much config required, as we do not really have any application-specific code. However, we'll lay the groundwork here. Open a terminal, and enter the following:

```
mkdir .vscode; \
cd .vscode; \
touch launch.json; \
touch settings.json; \
touch tasks.json; \
cd ..;
```

Add the following into `settings.json`:

```
{
    "thunder-client.saveToWorkspace": true,
    "thunder-client.workspaceRelativePath": ".thunder-client"
}
```

The preceding is some config for an HTTP testing tool that we will make use of in *Chapter 5*. We don't need to put anything into `launch.json` or `tasks.json` yet.

That should be all the configuration that is required for the time being, so we can move on.

Exiting the environment

Getting out of the dev container environment and back to your host is straightforward. Simply hit *F1* again and search for Reopen Folder Locally:

Figure 2.11 – Return to the local workspace

This will quickly return you to the host. You will no longer see the indicator in the lower-left corner that you are in a dev container, and the terminal will again connect directly to the host machine.

We have achieved a lot in this section, and we have set ourselves up to start working in a Dockerized environment. This may seem like a bit more work than simply starting to write code on your local machine, but I hope that as we proceed through this book together, you will start to see the value in taking the time to build out this environment right at the start of the project!

In the next section, we'll really start to build out the application and start to show the power of the preceding techniques.

Building out a sample application

In this section, we'll use the tools that we have just installed to create a very basic wireframe for a SaaS application. At this point, this will simply be a skeleton application that we can build out much more fully in subsequent chapters. However, this will introduce us to all of the different tools that we will be using.

The sample application that we will build as we work through this book will be an application to track a habit – something that you may like to try to do every day – such as learning a new language, or writing a few pages of a book in order to stay up to date with your publishing deadlines! I hope that this is something that you may be able to make use of after you have finished working through this book!

Database technologies

We will start with the database. We are using SQL Server as the database platform – this is a book based on Microsoft technologies, so we will stick with them as much as possible! However, other database platforms are available, and can easily be used. The Docker-based setup provided above makes it very easy to experiment with other database platforms. You can replace the SQL Server container with a Postgres container and see if everything still works!

The database and the database platform are often a source of some pain when getting developers onboarded and up to speed on a project. This is particularly evident in a SaaS project and is only partly mitigated using the Dockerized solution above.

It is very common these days to make use of an **Object Relational Mapper (ORM)** to manage the interactions between the code and the database, and this is a pattern that I will adhere to in this book. I will be using Entity Framework for all interactions initially, but note that when we discuss performance and scaling, I will touch on other techniques that can be used when database performance is paramount.

I will make use of Entity Framework's "code-first" approach to define the database and populate it with some initial startup data, and I'll make use of the migrations to keep the database up to date. This will come in very handy in later chapters when we talk about testing with a database, and also when we look at CI/CD and how to update a production database.

Creating the database

Make sure that you are in the dev container (check the bottom left of VSCode) and open a new terminal. Use the terminal to create a new .NET class library called GoodHabits.Database with the following code:

```
dotnet new classlib --name GoodHabits.Database;
```

You should see the folder appear in File Explorer, and the following output in the terminal:

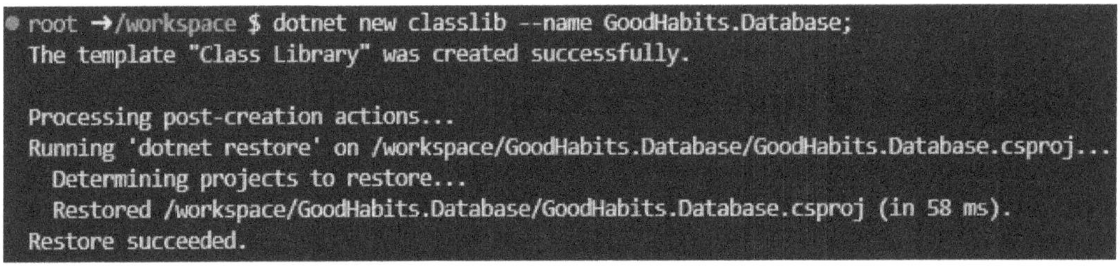

```
root →/workspace $ dotnet new classlib --name GoodHabits.Database;
The template "Class Library" was created successfully.

Processing post-creation actions...
Running 'dotnet restore' on /workspace/GoodHabits.Database/GoodHabits.Database.csproj...
  Determining projects to restore...
  Restored /workspace/GoodHabits.Database/GoodHabits.Database.csproj (in 58 ms).
Restore succeeded.
```

Figure 2.12 – Create the database

Before we can use this project to interact with the database, we need to add a few NuGet packages. So, again, in the terminal, enter the following:

```
cd GoodHabits.Database; \
dotnet add package Microsoft.EntityFrameworkCore; \
dotnet add package Microsoft.EntityFrameworkCore.Design; \
dotnet add package Microsoft.EntityFrameworkCore.Analyzers; \
dotnet add package Microsoft.EntityFrameworkCore.Relational; \
dotnet add package Microsoft.EntityFrameworkCore.SqlServer; \
dotnet add package Microsoft.EntityFrameworkCore.Tools; \
touch GoodHabitsDbContext.cs; \
rm Class1.cs; \
touch SeedData.cs; \
mkdir Entities; \
cd Entities; \
```

```
touch Habit.cs; \
cd ..;
```

With the preceding, we have instructed the .NET CLI to add all of the required NuGet packages to facilitate interactions with the database. We have also added a `GoodHabitsDbContext` class, a `SeedData` class, and a `Habit` class. We'll now add some basic setup into these three files that will give us a foundation to work on in later chapters.

Enter the following code into the `Habits.cs` file:

```
namespace GoodHabits.Database.Entities;
public class Habit
{
    public int Id { get; set; }
    public string Name { get; set; } = default!;
    public string Description { get; set; } = default!;
}
```

The above is a very straightforward entity class representing a habit that a user of the app may want to embed into their day-to-day lives.

Next, add some dummy data, by adding the following code to the `SeedData.cs` file:

```
using GoodHabits.Database.Entities;
using Microsoft.EntityFrameworkCore;

public static class SeedData
{
    public static void Seed(ModelBuilder modelBuilder)
    {
        modelBuilder.Entity<Habit>().HasData(
            new Habit { Id = 100, Name = "Learn French",
                Description = "Become a francophone" },
            new Habit { Id = 101, Name = "Run a marathon",
                Description = "Get really fit" },
            new Habit { Id = 102, Name = "Write every day",
                Description = "Finish your book project" }
        );
    }
}
```

Now create a `DbContext` by entering the following code into the `GoodHabitsDbContext.cs` file:

```
using GoodHabits.Database.Entities;
using Microsoft.EntityFrameworkCore;
namespace GoodHabits.Database;
public class GoodHabitsDbContext : DbContext
```

```
{
    public DbSet<Habit>? Habits { get; set; }
    protected override void
      OnConfiguring(DbContextOptionsBuilder options)
       => options.UseSqlServer("Server=
         sqlserver;Database=GoodHabitsDatabase;User
         Id=sa;Password=Password1 ;Integrated
         Security=false;TrustServerCertificate=true;");
    protected override void OnModelCreating(ModelBuilder
      modelBuilder) => SeedData.Seed(modelBuilder);
}
```

This does a few things. Firstly, we define a DbSet. This is mapped onto a table in the database.

Next, we hardcode the database connection string. Of course, it is bad practice to hardcode the connection string, and doubly bad to have the password there in plain text. We will correct these errors in *Chapter 3*, but this is sufficient to prove that we are connected to the database.

With this setup done, we can test this out and see whether we can migrate this information into the SQL Server database that we have running in a second Docker container.

To do this, let's start by using Entity Framework to create an initial migration. Enter the following into the terminal to generate the initial migration:

```
dotnet-ef migrations add InitialSetup;
```

You will see a Migrations folder has appeared in File Explorer with the InitialSetup migration modeled as a class. You don't have to worry too much about this at present, but it's worth taking the time to have a look at this class.

Then enter the following to deploy the migration to the SQL Server database:

```
dotnet-ef database update;
```

This sends the migration to the database.

And that is it for now. We have configured a basic database using Entity Framework in "code-first" mode, and have sent the first migration to the database.

How do we know that this has worked?

It's one thing to understand that the command has been executed successfully, but seeing is believing, and we need to dip into the database to really be sure that all of this is actually working as expected.

You will notice that when we defined the extensions that we wanted to be installed in the dev container, we specified the following extension should be included:

```
"ms-mssql.mssql",
```

This is an extension for VSCode from Microsoft that allows us to query a SQL server database directly from VSCode. Click on the extension and we will add a new connection, with the following information added:

Parameter	Value	Notes
Hostname	`sqlerver`	This is the name of the Docker container that we configured to run the SQL Server 2022 instance
Database to connect	`GoodHabitsDatabase`	This was defined in the `DbContext` class on the connection string
Authentication Type	`SQL Login`	
User name	`sa`	
Password	`Password1`	Defined in `sqlserver.env`
Save Password	`Yes`	
Display Name	`GoodHabits`	

You may need to okay a security popup.

With the above correctly entered, you should now get a view of the database, as follows:

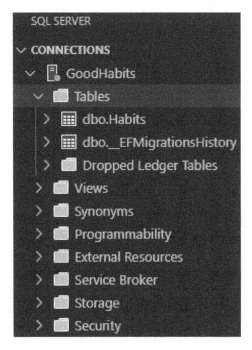

Figure 2.13 – Browse the database

You will note that the `Habits` table that we have defined as a `DbSet` in the `DbContext` file has been successfully migrated to the SQL Server database. You can right-click on the table named `dbo.Habits` and then click on **Select to 1000** to see the contents.

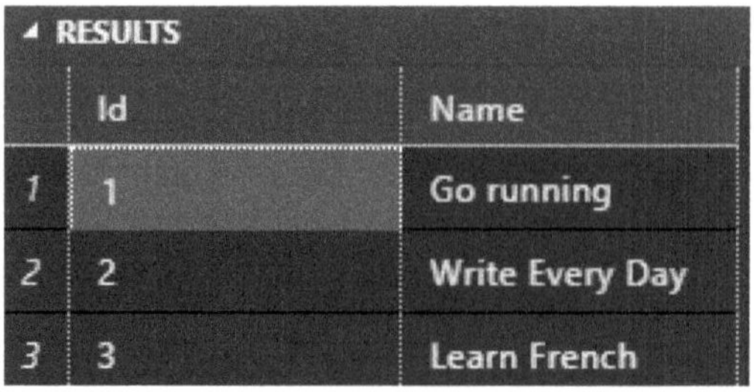

▲ RESULTS		
	Id	Name
1	1	Go running
2	2	Write Every Day
3	3	Learn French

Figure 2.14 – Data stored in the database

Again, you will see that the information that we have added in the `DbContext` file is present and correct in the database.

API technology

Next, we will move on to the API. The purpose of the API is to facilitate the storage and retrieval of information from the database by the users and also to do any required data processing.

There are many different technologies that are available with which to build an API. This book is focused on Microsoft technologies, so we will use the .NET Web API framework and the C# language. However, the tools and techniques described could easily be adapted to make use of different technologies.

Additionally, there are many different ideas about how best to structure an API, and there is no *one size fits all* approach that will work with every project out there. I have decided that I will use the RESTful paradigm for the examples in this book, but again, the concepts and ideas that are presented should copy over very well if your project is using some other structure, such as GraphQL.

The nature of a SaaS application is such that there are a huge number of choices to make when the project is getting started. This remains true even when the project is a demo application in a book!

Creating the HabitService API

Making sure that you are in the dev container environment, and are in the root of the project (the `workspace` folder), create a new `webapi` project with the following:

```
dotnet new webapi --name GoodHabits.HabitService; \
cd GoodHabits.HabitService; \
dotnet add reference ../GoodHabits.Database/GoodHabits.Database.
csproj; \
dotnet add package Microsoft.EntityFrameworkCore.Design; \
cd ..;
```

The .NET CLI does a lot of work setting up the API. We need to make a change to the `launchSettings.json` file. Open the `Properties` folder and replace the default launch settings with the following:

```
{
  "$schema":
    "https://json.schemastore.org/launchsettings.json",
  "profiles": {
    "HabitService": {
      "commandName": "Project",
      "dotnetRunMessages": true,
      "launchBrowser": false,
      "applicationUrl": "http://localhost:5100",
      "environmentVariables": {
        "ASPNETCORE_ENVIRONMENT": "Development"
      }
    }
  }
}
```

The most important thing to note is that we will be running the `HabitService` on port `5100`. This is important to keep track of when we start to look at microservices in *Chapter 6*.

Out of the box, the default web API comes with the Weather API, which you can take a look at to get an idea of how these endpoints are configured. We can test this by entering the following into the terminal:

```
dotnet run
```

This will start the application running on the port that is specified in `launchSettings.json` – port `5100` in this case.

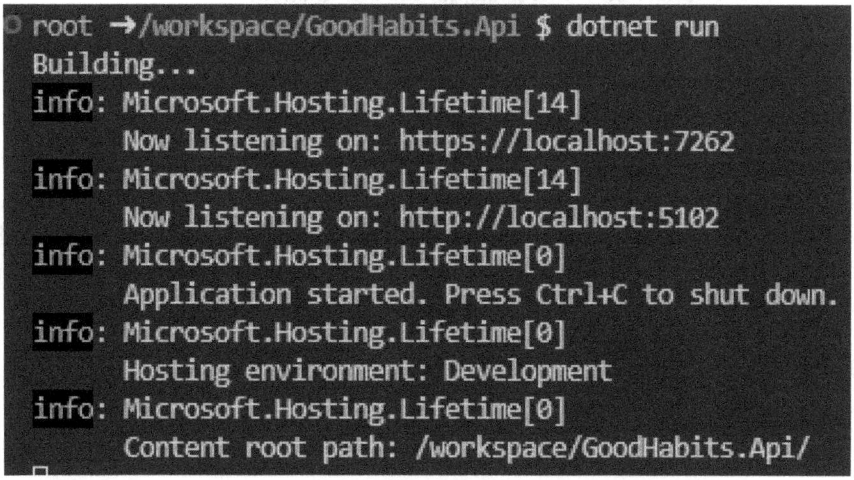

```
root →/workspace/GoodHabits.Api $ dotnet run
Building...
info: Microsoft.Hosting.Lifetime[14]
      Now listening on: https://localhost:7262
info: Microsoft.Hosting.Lifetime[14]
      Now listening on: http://localhost:5102
info: Microsoft.Hosting.Lifetime[0]
      Application started. Press Ctrl+C to shut down.
info: Microsoft.Hosting.Lifetime[0]
      Hosting environment: Development
info: Microsoft.Hosting.Lifetime[0]
      Content root path: /workspace/GoodHabits.Api/
```

Figure 2.15 – Console output indicating success

You can check to see whether the application is running by going to the following URL in your browser (remember to check the port number!): `http://127.0.0.1:5100/swagger/index.html`

Note that you may get some HTTPS warnings from your browser, as follows:

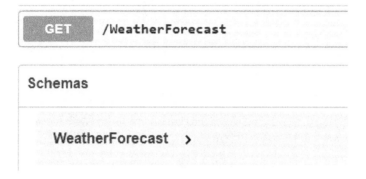

Figure 2.16 – The GoodHabits API

User interface technology

Finally, we need a user interface on which to display the information.

As I have pointed out for both the database and the API, there are many different UI technologies, and the underlying tools and techniques that you will learn in this book can be applied to any one of them. Often, the best technology to use in any given situation is the one that you are most comfortable with.

This is a Microsoft-centric book aimed at existing .NET developers, so I will use Blazor for the examples that are provided. If you prefer Angular, React, Vue, or any of the other millions of UI frameworks that are out there, please don't be put off. In fact, adapting these examples to work with your UI of choice would be an excellent exercise to further your understanding of the concepts that we will cover in this book.

Creating the UI

Creating a simple demo Blazor UI is straightforward with the CLI tools provided by Microsoft. Enter the following commands into the terminal to create the UI:

```
dotnet new blazorwasm -o GoodHabits.Client;
```

This follows the same pattern that we used to create the HabitServer project. And, like the HabitServer project, we will need to modify the launch configuration in `launchSettings.json`. Again, look in the `Properties` folder, and overwrite the contents with this:

```
{
  "profiles": {
    "GoodHabitsClient": {
      "commandName": "Project",
      "dotnetRunMessages": true,
      "launchBrowser": true,
      "inspectUri": "{wsProtocol}://{url.hostname}:
        {url.port}/_framework/debug/ws-
        proxy?browser={browserInspectUri}",
      "applicationUrl": "http://localhost:5900",
      "environmentVariables": {
        "ASPNETCORE_ENVIRONMENT": "Development"
      }
    }
  }
}
```

Again, take note of the port that the client is running on. We will use `5900` for the client.

With the config done, you can now start the client by typing `dotnet run` into the console.

Again, you will see that the `Client` app is running on the ports specified in `launchSettings.json`, which should be port `5900`:

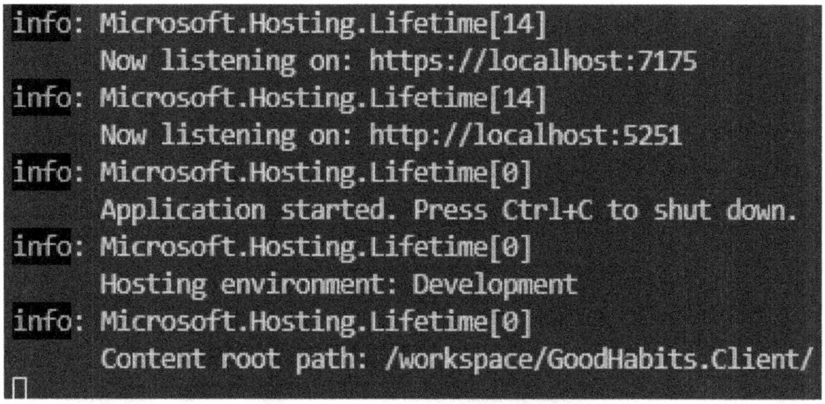

```
info: Microsoft.Hosting.Lifetime[14]
      Now listening on: https://localhost:7175
info: Microsoft.Hosting.Lifetime[14]
      Now listening on: http://localhost:5251
info: Microsoft.Hosting.Lifetime[0]
      Application started. Press Ctrl+C to shut down.
info: Microsoft.Hosting.Lifetime[0]
      Hosting environment: Development
info: Microsoft.Hosting.Lifetime[0]
      Content root path: /workspace/GoodHabits.Client/
```

Figure 2.17 – Console output indicating success

Again, as per the API, you can follow this link in your browser to see the Blazor app running:

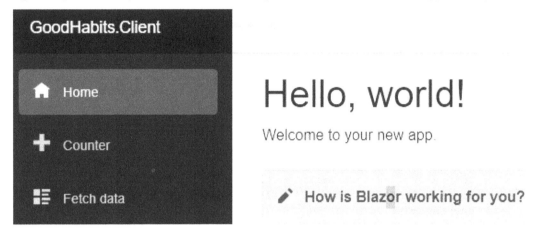

Figure 2.18 – Hello, world!

Starting the application

Currently, we only have two projects to run, the HabitService API and the Blazor client. So, we can fairly easily start the whole project by typing `dotnet run` twice. As we build on this application, it will become more and more complex, and harder to run in this way. So, we'll create build and launch configurations to tell VSCode how to start the application.

We have already created the config files for this in the `.vscode` folder.

Start by opening `tasks.json`, and copy in the following:

```
{
    "version": "2.0.0",
    "tasks": [
        {
            "label": "build-client",
            "type": "shell",
            "command": "dotnet",
            "args": [
                "build",
                "${workspaceFolder}/GoodHabits.Client/
                    GoodHabits.Client.csproj"
            ],
            "group": {
                "kind": "build",
                "isDefault": true
            }
```

```
        },
        {
            "label": "build-habit-service",
            "type": "shell",
            "command": "dotnet",
            "args": [
                "build",
                "${workspaceFolder}/GoodHabits.HabitService
                    /GoodHabits.HabitService.csproj"
            ],
            "group": {
                "kind": "build",
                "isDefault": true
            }
        },
    ]
}
```

You can see in the above JSON that two tasks are defined to build the client and the habit service.

Next, modify launch.json by adding the following JSON config:

```
{
    "version": "0.2.0",
    "configurations": [
        {
            "name": "RunClient",
            "type": "blazorwasm",
            "request": "launch",
            "preLaunchTask": "build-client",
            "cwd": "${workspaceFolder}/GoodHabits.Client",
            "url": "https://localhost:5900"
        },
        {

            "name": "RunHabitService",
            "type": "coreclr",
            "request": "launch",
            "preLaunchTask": "build-habit-service",
            "program": "${workspaceFolder}/
                GoodHabits.HabitService/bin/Debug/net7.0/
                GoodHabits.HabitService.dll",
            "args": [],
            "cwd":
                "${workspaceFolder}/GoodHabits.HabitService",
            "stopAtEntry": false,
            "console": "integratedTerminal"
```

```
            },
        ],
        "compounds": [
            {
                "name": "Run All",
                "configurations": [
                    "RunHabitService",
                    "RunClient"
                ]
            }
        ]
    }
```

Again, you can see that two configurations are added to run the individual project. You can also see nearer the bottom that there is a `compounds launch` command set to run all of the projects.

You can test this out by pressing *Ctrl + Shift + D* to enter the **Run and Debug** menu, selecting **Run All** from the dropdown, and pressing the play button.

You will see that this starts the API and the client at the same time. This will be very handy as the number of projects starts to grow!

Note that you can also hit *F5* to start the application as well.

Adding a solution file

There is one last small piece of setup to do before we move on to really building out the application, and that is to add a solution file. This is not strictly necessary, but it is commonly done when working with .NET projects and will allow us to easily build, clean, and test the projects with a single command.

To add a solution file, simply navigate to the project root, and run the following:

```
dotnet new sln --name GoodHabits; \
dotnet sln add ./GoodHabits.Client/GoodHabits.Client.csproj; \
dotnet sln add ./GoodHabits.HabitService/GoodHabits.HabitService.
csproj; \
dotnet sln add ./GoodHabits.Database/GoodHabits.Database.csproj;
```

This simply adds a solution file and references the three projects that we have created.

And that is the final piece of setup work to do – now we can progress to building the application.

Where do we go from here?

We have done a lot of work in this chapter, but we haven't really started to build the application yet. We have instead focused on choosing the tools that we will use and building a development environment around those tools. We now have the outline of a SaaS application that we can continue to work with as we move through this book.

Taking the time at the start of the project to select the correct tools is an important step of any SaaS project, and it should not be skipped.

Being successful in developing any application requires that some thought is put into the technologies and frameworks that will be used, and also the tooling that will be used. In this chapter, we have set ourselves up very well to be able to get our hands dirty and try things out as we explore the concepts that are required to build a SaaS application.

The following chapters will start to introduce a lot more SaaS-specific considerations, and we will use this outline app to demonstrate them.

Summary

In this chapter, we have briefly touched on a lot of different tools, topics, and techniques. This is the nature of developing SaaS applications – please don't feel overwhelmed! To get started, we installed the tools that we need to work with, namely Docker Desktop and Visual Studio Code. This is quite a light set of tools for a SaaS project. But as you have seen, we used Docker to wrap up the developer environment. We learned about dev containers, and how they significantly streamline project setup for complex projects, such as a SaaS application, and we then configured a dev container and learned how to work in that environment.

With the core of the environment set up, we set up a database and populated it with some data using Entity Framework, then made the data available through an API, and finally surfaced the data on a Blazor UI!

With all of the above, we have configured all of the individual parts required to build up a fully featured SaaS application. Read on, and we will do just that!

In the next chapter, you will learn about a core part of building SaaS applications, and that is multi-tenancy. We will cover what this is and why it is so important, and start to think about how we will implement it in our sample application.

Further reading

- Developing inside a Container: `https://code.visualstudio.com/docs/devcontainers/containers`

- Entity Framework Dev Environment in Docker: `https://itnext.io/database-development-in-docker-with-entity-framework-core-95772714626f`

- .NET Development in Docker with Dev Containers: `https://itnext.io/net-development-in-docker-6509d8a5077b`

- Blazor Tutorial - Build your first Blazor app: `https://dotnet.microsoft.com/en-us/learn/aspnet/blazor-tutorial/intro`

- Tutorial: Create a web API with ASP.NET Core: `https://learn.microsoft.com/en-us/aspnet/core/tutorials/first-web-api?view=aspnetcore-7.0&tabs=visual-studio`

Questions

1. What are the advantages of using dev containers?

2. How were we able to use SQL Server without having to install it on our developer machines?

3. What are the HTTP verbs commonly associated with RESTful APIs?

4. What are the benefits of using Entity Framework?

We have covered a lot in this chapter! Don't worry if the answers to the above questions are not 100% clear at this stage – we will expand on all of them in the coming chapters.

Part 2:
Building the Backend

This section covers all things backend related, starting with the database and building up to the API layer. This section also introduces the topics of multi-tenancy and microservices, both of which are very important for developing SaaS applications.

This section has the following chapters:

- *Chapter 3, What Is Multi-Tenancy, and Why Is It Important in SaaS Applications?*
- *Chapter 4, Building Databases and Planning for Data-Rich Applications*
- *Chapter 5, Building RESTful APIs*
- *Chapter 6, Microservices for SaaS Applications*

What Is Multi-Tenancy, and Why Is It Important in SaaS Applications?

Multi-tenancy has become a critical architectural pattern in modern **Software as a Service** (**SaaS**) applications, enabling providers to serve multiple customers (tenants) from a single instance of their software.

This chapter will delve into this very important part of building SaaS applications –termed *multi-tenancy* – whereby multiple tenants can use the *same instance* of a deployed application and still keep their data private and segregated.

Under a multi-tenancy system, multiple customers of the SaaS application can use the same instance of the application and also optionally use the same database, but their data is segregated, such that no other tenant can see the data – unless it is explicitly shared. This obviously raises a number of concerns about security and data privacy. For a company building a multi-tenant SaaS application, it is extremely important to ensure that any given customer is only ever able to see their own data and never anyone else's!

Many strategies, patterns, utilities, and technologies can be leveraged to ensure that an individual tenant's data remains segregated, but the first line of defense is – as always –a sound technical understanding of the underlying concepts. It is particularly important that the development team appreciates this when the application is being crafted. Multi-tenancy is essential for any SaaS application that will scale, which almost all of them will aim to do!

The privacy and security of data in a SaaS multi-tenant application are paramount. It's hard to think of any instance where a multi-tenant SaaS application is built that did not require at least some of the tenants' and users' data to remain private. Robust data security and segregation of data are, therefore, key considerations when setting out to build a SaaS application.

When starting out on the development cycle for a new application, it is considerably easier to build robust security into the application at the very start of the project than it is to retrospectively add it in later! It is also much less risky to get the security and segregation working correctly from the start – once the inevitable application sprawl and feature creep start, it is considerably harder to convince yourself that everything really is locked down as tightly as it should be.

We are covering this topic early in this book so that, as we build out our demo application, you will have a solid understanding of multi-tenancy and how it impacts future design choices. I strongly encourage you to take the same approach when building a SaaS application, taking the time at the start of the project to plan a strategy for multi-tenancy and the security considerations that will arise.

During this chapter, we will cover the following main topics:

- Gaining an understanding of what multi-tenancy is

- Considering the options for data storage with multi-tenant applications

- Understanding the design considerations through the application layers

- Discussing security considerations

By the end of the chapter, you'll have a good understanding of multi-tenancy, multi-tenant applications, and the specific security considerations that will arise when working on SaaS and, specifically, multi-tenant software applications.

Technical requirements

All code from this chapter can be found at `https://github.com/PacktPublishing/Building-Modern-SaaS-Applications-with-C-and-.NET/tree/main/Chapter-3`.

What is multi-tenancy?

Multi-tenancy is a software architecture pattern commonly used in SaaS applications, wherein a single application instance serves multiple customer groups or organizations, known as tenants. Each tenant shares the same underlying infrastructure and resources, such as servers, databases, and application logic, but maintains its own isolated data and configurations. This approach allows for optimized resource usage, streamlined maintenance, and reduced operational costs. The primary challenge in a multi-tenant architecture is to ensure data isolation, security, and performance for each tenant, while also offering customization and scalability. In essence, multi-tenancy enables SaaS providers to deliver a cost-effective, secure, and customizable solution to a diverse range of customers, using a shared application instance.

Disambiguating customers, tenants, and users

When talking about multi-tenancy, is it important to disambiguate the related but distinct concepts:

- **Customers**: Customers refer to the organizations or individuals who purchase or subscribe to a software product or service. In the context of SaaS applications, customers may be represented as one or more tenants, with each customer group having access to its own isolated environment within the shared application instance. The term "customer" generally emphasizes the business relationship and financial aspect of using a software product or service.

- **Tenants**: Tenants represent separate organizations, companies, or groups that use a shared instance of a SaaS application in a multi-tenant architecture. Each tenant has its own isolated data, configurations, and customizations while sharing the same software infrastructure and resources with other tenants. A key aspect of a multi-tenant system is that each tenant can have multiple users associated with it, enabling individual access and personalized experiences within the shared application instance.

- **Users**: Users are the individual people who interact with a software system or application, often with unique login credentials and personalized settings. Users belong to a specific tenant, organization, or group, and their actions and access within the system can be managed with role-based permissions or access control. Users represent the actual individuals using the software, carrying out various tasks and activities within the application.

Let's consider an example where two large companies use a SaaS application. The companies are customers, and to keep this example simple, each will have a single tenant in the application. Each company could have hundreds or thousands of employees who use the system. It would be typical for both companies to want to customize the app in some way for their own employees. For example, they may want to display their logo when a user logs in. This would be achieved by having two *tenants* in the application, each configured to show the correct logo to the users who belong to that company.

There will be two levels of data segregation. The individual tenant configuration (the logo, for example) will be separate, and any user operating within that tenant must always see the correct logo. There will be a further level of segregation, keeping each individual user's data private.

Users with elevated permissions within a tenant (admin users) may be able to modify details for other users within that same tenant, but never for any other tenants.

It is very important to keep this in mind as we progress through this chapter – make sure that you understand the difference between users and tenants before moving on:

- A customer can have one or more tenants in an application, often one, meaning that the terms *customer* and *tenant* can be used interchangeably

- A tenant can have one or more users under it, most often more than one, meaning that the terms should not be used interchangeably

What are the alternatives to multi-tenancy?

While SaaS applications are largely synonymous with multi-tenant applications, it is technically possible to deliver a SaaS application as a single-tenant application.

If a single-tenant approach is followed, then every time a new tenant is onboarded, there is an entirely new application stack and a new database deployed for that specific tenant. There *could* be some very specific and limited shared services, such as a shared login page that redirects the customer to their own instance of the application, but in general, this will be a completely unique and isolated deployment of the software on a per-customer basis.

A single-tenant application is generally considered to be the most secure method and could be considered the most reliable method to deliver an application to a single customer. However, it is significantly most expensive to do this for each and every customer, and the costs to scale such an architecture quickly become unmanageable.

The resources required to maintain, support, upgrade, and customize such an application are very high, meaning that the price to the end user is also high, and would typically restrict your customer base to the enterprise. If your end goal is to attract non-enterprise customers, this single-tenant approach is very unlikely to be successful.

It is also somewhat stretching the definition of SaaS. If each customer simply gets a unique installation of the software, albeit in the cloud, then it is much closer to the traditional method of delivering software – on a CD with a unique key printed on it!

Single tenancy in a SaaS application would really only be done for a very small subset of use cases. For this book, we will not consider this any further, other than to mention that it is technically possible to build a SaaS application without making use of a multi-tenant architecture!

The advantages and disadvantages of multi-tenancy

Some of the advantages of using a multi-tenant architecture are as follows:

- **Cost**: Multi-tenancy is typically a very cost-efficient way to deliver an application from the point of view of provisioning resources (app servers, databases, and so on). Multi-tenancy also tends to be very cost-efficient when considering ongoing support and maintenance costs. The marginal cost for every additional tenant, and each additional user within that tenant, will typically not add much to the overall cost once the application has gone live.

- **Tiered pricing**: Customers can be offered a range of pricing options to suit them and their organization specifically. The pricing can scale linearly as new tenants are added and users are introduced from the new tenant's organization.

- **Easy to update**: With only one instance of the application being accessed by many tenants, a single update can be performed that will get every user across all tenants on the latest and greatest version. Note that it is also possible, if more complex, to configure this using **Continuous Deployment (CD) pipelines** under a single-tenant model.

- **Scalability**: A SaaS application using a multi-tenant architecture will typically scale up very easily as the number of tenants and users increases. Assuming that a cloud provider is used, this can happen literally effortlessly on the part of the development team. A cloud service can be configured so that as additional demand is placed on the system by the growing user base, additional resources are deployed automatically.

It's not all good, though – there are a few disadvantages that should also be considered:

- **Increased complexity**: From a software architecture point of view, a multi-tenant application is almost by definition more challenging to build. There are additional complexities throughout the application stack, starting with the database requiring some form of segmentation by tenant all the way up the stack to the user interface, where each user must be securely authenticated and authorized to access only certain parts of the system.

- **Increased security requirements**: The fact that multiple tenants share a single application instance necessitates a much more thorough approach to user security. It is typically an absolute disaster scenario for a business if one of their user's private data is leaked to another user through their application, and doubly so if that user belonged to a different tenant organization.

- **Downtime is a disaster**: If a multi-tenant SaaS system goes down for any reason, then typically, every single customer will have no access to the application. This obviously makes reducing downtime absolutely critical.

- **Noisy neighbors**: Given that each tenant shares an application, they are, therefore, also sharing resources, such as the compute time of the cluster or the servers on which the application is deployed. One particularly compute-hungry user could have a knock-on effect on every other user of a system.

An observation from the preceding lists is that the advantages provided are significant and cannot really be worked around if using a single-tenant architecture. However, the disadvantages of a multi-tenant system can usually be mitigated by simply taking more time upfront to design the system well. Of course, there is a cost in doing so, which must be balanced carefully when choosing the application architecture and making decisions about system design.

In general, though, the higher upfront cost will be returned many times over as the number of tenants and the user base starts to grow and the power of the SaaS/multi-tenant application is realized!

I hope that with the preceding discussion, I have convinced you of the benefits of multi-tenant architecture and encouraged you to consider this at the very start of your project. We can now move on to discuss the specifics of designing for multi-tenancy, starting with the most important – data storage!

Options for data storage with multi-tenant applications

A database in any application is typically the foundation upon which the rest of the application is built. The decisions that are taken when selecting and designing the database will have significant knock-on effects on the data access/API layer and probably some impact on the user interface as well.

Additionally, the data stored in an application will represent your customers' business information. Often, this information will be incredibly valuable to them, and by entrusting it to you – as a developer working on a SaaS application – they are showing significant trust that you can keep this valuable asset safe. The database underpinning a SaaS application should be like a bank vault where your customers are happy to deposit their data!

Therefore, making the correct choices in database design is extremely important for the rest of the application development, and also for the tenants and individual users of the application.

While this section is focused on the database, its role as the foundation for the rest of the application will necessitate some discussion about how the choices made will impact other layers and parts of the application.

Key considerations

The approach taken for data storage is very important, with a few key areas to consider before you start building.

Design complexity

Generally, when starting to build a new application, it is a good idea to focus on simplicity in design, and only add complexity when it is required. Building in multi-tenancy does add significant complexity to the application, so it is very important that the level of complexity is considered and the solution is sized accordingly.

Consider how many tenants you expect to have, and pick a solution that matches that. Consider also how important data isolation is, and pick a solution that matches that.

You should also keep in mind that some customers – particularly large corporate customers – may expect some degree of customization. Allowing per-customer customizations can start to exponentially increase the complexity of a design. While it is generally preferred to avoid any kind of customization, this may not be possible. If it is felt that this will be a requirement, then some consideration for this eventuality should be taken into account at the design stage.

Support complexity

While the process of developing the application is, of course, costly for the company that is funding the undertaking, it is important to consider the ongoing cost of actually running the application. This is the period in the application life cycle where it is expected to generate revenue, and that will not be possible if there are huge ongoing support and maintenance costs.

One very important aspect of ongoing maintenance is monitoring your tenants' usage. The "80–20" principle will most likely apply, and you need to know which of your tenants are the most profitable… and which are the most problematic!

You need to consider that some of your tenants will have higher availability requirements than others. Will you be able to support a tenant with a 99.99% uptime requirement? And do you understand the additional costs and technical trade-offs in offering this level of service?

Support and maintenance can quickly become a technical and financial nightmare if they are not considered early on in the project life cycle.

Scale

No application ever launched with a billion users – they all started with one!

When planning a SaaS application, you need to have at least some idea of the scale that you expect to operate at over the short, medium, and long term. With this understanding, you can make smart decisions that suit the individual users and the tenants from launch day through to platform maturity.

There is no point in wasting effort and resources to build for a billion users on day one. Similarly, there is no way to get to a billion users if there is no plan in place to adequately service that number of users.

Performance

Under a multi-tenant system, many users will be sharing – and, therefore, competing for access to – resources. This includes users under separate tenants, who would typically work for different organizations.

You need to consider how a typical tenant and a typical user will use the system. For example, if you are building an enterprise system and focusing on a single time zone, you should expect to see almost all the usage during business hours. You will need to size your resources to meet this demand, even though they will be idle out of business hours.

You may face a similar problem with – for example – a streaming service that will spike in usage in the evening.

Note that this scenario would offer a considerable advantage to a company offering its services across multiple time zones, where the usage spikes could be flattened out.

Users in a multi-tenant system competing for shared resources is known as the "noisy neighbor" problem. This phenomenon is where one tenant or user is exhibiting particularly heavy resource usage and, by doing so, degrades the system performance for its other users. This problem is inevitable to a certain extent when building a multi-tenant system, but there are ways it can be mitigated, such as the following:

- Implementing throttling to prevent a single tenant or user from consuming a disproportionate amount of resources.

- Monitoring how the tenants and users interact with the system and working around the more resource-intensive examples. For example, they could be moved to a separate cluster.

- Purchasing additional cloud resources. This is something of a blunt instrument, but it is effective.

Isolation

As we h discussed previously, it is of the utmost importance that data that relates to one tenant is invisible to another tenant, and each of the users within those tenants. An individual tenant's data must be isolated. There are a few ways this can be achieved:

- Containers can be used on a one-container-per-tenant basis. This can be a very effective method to compartmentalize tenants of a system, but there would be scaling concerns should the application turn into the next Facebook.

- Separate tables can be used per tenant in the same database. This provides a good degree of assurance that data cannot "leak," but again, this would not scale efficiently to hundreds of tenants and millions of users.

- Tenant ID-based isolation, where data in the database is all in one table, with a `tenant_id` column. This scales very well but could be considered to be less secure than the previous options.

There is no "one-size-fits-all" approach to this problem. The level of isolation required will depend on the type of customers and the type of data being stored. This should be carefully considered at the start of a project, as changing the approach to data isolation later in the project life cycle can be extremely challenging.

Cost

Of course, we would like the most secure, performant, and easy-to-maintain application possible, but as application developers, we must always operate within the confines of the economic realities that are faced by the company building the application.

Even with matters as important as data security, there is always a cost trade-off to be made.

It is extremely helpful if these are understood at the start of a project so that there are no nasty surprises later!

A database is foundational to the rest of an application, and making the correct choices regarding the aforementioned considerations is extremely important. The choices that are made at this stage will significantly impact the rest of the application. We will consider this in the next section.

Design considerations through the application layers

The approach that will be taken to solve a multi-tenancy problem is probably one of the most important decisions to take in the early stages of a project. It can be one of the most expensive and technically challenging aspects to make a change to later in the project life cycle. The planning and the design of a solution for the multi-tenant problem must start in the very early phases and will influence all of the various layers in the application.

The core reason for multi-tenancy is to segregate data that one customer stores such that it is not visible to a user under a different tenant.

The key decision to make is where the data will be segregated, and this primarily affects the design of the database. The decision driver for this will have to be related to the users of a system. To make this decision, it is vital to understand the following:

- Who your tenants will be.

- How your users will use a system.

- What the consequences of a data breach are to your customers.

- What the customers' tolerance for risk versus cost is. Would they prefer a more costly but more secure solution or a cheaper solution?

- What the consequences of a data leak are to a business

Users of social media sites do not seem overly concerned for their privacy, and so will likely tolerate a less security-conscious approach to data segregation. Also, users of social media are often not prepared to pay anything for the system, so they are very price-conscious and not at all data privacy-conscious. Indeed, the business model of a social network is often to sell users' data to the highest bidder!

Enterprise customers will almost never tolerate a social media-esque approach to data and will always prefer a more secure regime. Some enterprise customers will not accept a multi-tenant system at all. Military or government customers, for example, would almost certainly consider the risks too high and would insist on a single-tenant solution.

If the customers of an application are likely to be primarily motivated by security, and in particular, rigorous data segregation and an absolute guarantee that there will be no slip-ups, then the best solution is to not design a multi-tenant application and, therefore, not build a SaaS application! Given that this book is about developing SaaS applications, we will discount this customer class entirely. However, it is important to be aware of the limitation of the technique. A customer base that is primarily motivated by absolute data security over all other considerations is very unlikely to accept an app that is delivered via SaaS. These users will most likely require on-premises solutions hosted inside their own network, with full control over the data retained by the users' organization.

Given that we are building for customers that we feel will buy a SaaS application, let's consider those!

One database per tenant

The most secure way to segregate one tenant's data from another is to store the data in completely separate databases. This has a considerable overhead in terms of maintaining and updating the multiple databases (one per customer) that are required under this scheme. This approach all but guarantees there will be no data leaks, but the additional overheads are considerable. An application that uses this scheme will also present challenges in scaling, and it may lead to spiraling costs if there is a per-instance license for the database platform that is selected.

The issue of scaling does present a hard cap on the number of tenants that a system could be designed for. There is no way that Microsoft could make their DevOps platform work if every tenant organization had a unique and segregated database – there are simply too many customers for that to be feasible.

Aside from security, one other benefit of this scheme is that the number of rows of data stored in the individual database will be lower than in a shared database, so there may be a slight performance boost possible in the database layer. Note that this may be offset in the application layers.

This type of scheme would be used only if the number of customers was quite small and were hugely security-conscious. There would be an argument in these cases to build single-tenant, on-premises systems.

Shared schema

This is the most common solution that is used for the vast majority of commonly used SaaS applications that you may encounter in day-to-day use. Even some potentially sensitive applications, such as a system that hosts financial and taxation information, may make use of a shared schema.

Under this regime, a single database is used, with tables containing data from multiple tenants. The data is protected from being shared using some form of identification in the database.

Using this scheme significantly boosts the benefits that are derived from using the SaaS paradigm in the first place. These days, there are plenty of well-understood and well-used ways of keeping data secure under such a scheme. There may also be an argument to be made that having only one database and one schema to keep on top of actually makes it easy to secure a system.

One schema per tenant

This is the halfway house between the two aforementioned options. Under this method, a single database is used, but the same schema is deployed to the database multiple times, once per customer. Each schema is isolated, and there are no connections or relations between the individual customers' schema.

In some ways, this is the best of both worlds, and in other ways, it's the worst.

There is still a significant additional overhead in managing the many schemas, and there is still a hard cap on the number of individual tenants that this could be scaled up to. Particularly security-conscious customers may still be turned off by this, considering it too much of a risk to even share a database.

One table per tenant

This is very closely related to the aforementioned *one schema per tenant* method. Under this paradigm, every tenant in the system has a table added to the database that pertains only to them.

The pros and cons of the previous method also apply here.

Examples

To illustrate these points, I'll build an example that demonstrates the two preceding segregations.

We will imagine there are four tenant organizations of the GoodHabits app. These tenants are as follows:

- AscendTech, who is very security-conscious and wants a private database
- Bluewave, who is also security-minded and also wants a private database
- CloudSphere, who is a price-conscious customer and takes the cheaper option to have a shared database
- DataStream, who will also use the shared database and share it with CloudSphere

With these four example organizations, we can demonstrate how a multi-tenant database can work. This example will not be exhaustive but will provide a solid foundational understanding.

In this section, we will build on the GoodHabits database and API project that we created in the previous chapter.

Adding the required packages

We only need to add one nuget package to the HabitService project for this chapter. Navigate to the project folder in a terminal, and enter the following:

```
dotnet add package Microsoft.AspNetCore.Mvc.NewtonsoftJson;
```

Modifying the entity

Many of the tables in the database will require a column that identifies the owner of the data – the tenant! To facilitate this, we can start by creating an interface called IHasTenant. Create the file with the following commands:

```
cd GoodHabits.Database; \
mkdir Interfaces; \
cd Interfaces; \
touch IHasTenant.cs; \
cd ..;
```

Then, copy the following code into the file:

```
namespace GoodHabits.Database;
public interface IHasTenant
{ public string TenantName { get; set; } }
```

You will recall that we previously created an entity class called `Habit` that defined the data structure for the database. This entity will need to have a `TenantName` value, so it should implement the `IHasTenant` interface. Modify the class that we created in *Chapter 2* such that it looks like this:

```
namespace GoodHabits.Database.Entities;
public class Habit : IHasTenant
{
  public int Id { get; set; }
  public string Name { get; set; } = default!;
  public string Description { get; set; } = default!;
  public string TenantName { get; set; } = default!;
}
```

Creating the tenant configuration

The configuration for the tenants in a live system would typically be stored in a separate database somewhere – a client portal application or similar. However, for the purposes of this demo, we will just use the `appsettings.json` file. Add the following to the `appsettings.json` file in the `Goodhabits.HabitService` project:

```
{
  "Logging": {
    "LogLevel": {
      "Default": "Information",
      "Microsoft.AspNetCore": "Warning"
    }
  },
  "AllowedHosts": "*",
  "TenantSettings": {
    "DefaultConnectionString": "Data
      Source=sqlserver;Initial
      Catalog=GoodHabitsDatabase;
      User Id=sa;Password=Password1;
      MultipleActiveResultSets=True;
      TrustServerCertificate=True;",
    "Tenants": [
      {
        "TenantName": "AscendTech",
        "ConnectionString": "Data Source=sqlserver;
          Initial Catalog=AscendTechGoodHabitsDatabase;
          User Id=sa;Password=Password1;
          MultipleActiveResultSets=True;
          TrustServerCertificate=True;"
      },
      {
        "TenantName": "Bluewave",
```

```
         "ConnectionString": "Data Source=sqlserver;
         Initial Catalog=BluewaveGoodHabitsDatabase;
         User Id=sa;Password=Password1;
         MultipleActiveResultSets=True;
         TrustServerCertificate=True;"
      },
      {
         "TenantName": "CloudSphere"
      },
      {
         "TenantName": "Datastream"
      }
    ]
  }
}
```

The preceding configuration has defined four tenants:

- The first two, AscendTech and Bluewave, want fully segregated databases and have had connection strings specified that are unique to them

- The second two, CloudSphere and Datastream, do not have a unique connection string, so they are considered to be happy to use the shared database

In the `GoodHabits.Database` project, add a class that matches the structure of the config that we added into `appsettings.json` previously. The config will be loaded into this at startup. Call the class `TenantSettings`, and then paste in the following:

```
namespace GoodHabits.Database;

public class TenantSettings
{
    public string? DefaultConnectionString { get; set; }
    public List<Tenant>? Tenants { get; set; }
}
public class Tenant
{
    public string? TenantName { get; set; }
    public string? ConnectionString { get; set; }
}
```

Creating a tenant service

Next, we will create a service that can provide details about the tenants to other parts of the application that may be interested.

Start in the `GoodHabits.Database` project by adding an interface called `ITenantService`, and then add the following code:

```
namespace GoodHabits.Database;
public interface ITenantService
{
    public string GetConnectionString();
    public Tenant GetTenant();
}
```

Next, we need to implement this service. This is done in the HabitService project and should look like this. Double-check that you are adding this to the HabitService project and not the Database project:

```
using Microsoft.Extensions.Options;
using GoodHabits.Database;
namespace GoodHabits.HabitService;
public class TenantService : ITenantService
{
    private readonly TenantSettings _tenantSettings;
    private HttpContext _httpContext;
    private Tenant _tenant;
    public TenantService(IOptions<TenantSettings>
      tenantSettings, IHttpContextAccessor contextAccessor)
    {
        _tenantSettings = tenantSettings.Value;
        _httpContext = contextAccessor.HttpContext!;
        if (_httpContext != null)
        {
            if (_httpContext.Request.Headers.TryGetValue(
              "tenant", out var tenantId))
            {
                SetTenant(tenantId!);
            }
            else
            {
                throw new Exception("Invalid Tenant!");
            }
        }
    }
    private void SetTenant(string tenantId)
    {
        _tenant = _tenantSettings!.Tenants.Where(a =>
          a.TenantName == tenantId).FirstOrDefault();
        if (_tenant == null) throw new Exception("Invalid
          Tenant!");
        if (string.IsNullOrEmpty(_tenant.ConnectionString))
```

```
                SetDefaultConnectionStringToCurrentTenant();
        }
        private void
            SetDefaultConnectionStringToCurrentTenant() =>
            _tenant.ConnectionString =
                _tenantSettings.DefaultConnectionString;
        public string GetConnectionString() =>
            _tenant?.ConnectionString!;
        public Tenant GetTenant() => _tenant;
    }
```

The primary function of the preceding code block is to intercept incoming HTTP requests, check that there is a tenant named in the header, and match that name with a known tenant.

Modifying the SeedData and AppDbContext classes

You will remember that in *Chapter 2*, we added some seed data. As we now require a tenant name to be present in the database, we will have to update the seed data. Copy in the following, or just add the `TenantName`:

```
using GoodHabits.Database.Entities;
using Microsoft.EntityFrameworkCore;

public static class SeedData
{
    public static void Seed(ModelBuilder modelBuilder)
    {
        modelBuilder.Entity<Habit>().HasData(
            new Habit { Id = 100, Name = "Learn French",
            Description = "Become a francophone",
            TenantName = "CloudSphere" },
            new Habit { Id = 101, Name = "Run a marathon",
              Description = "Get really fit",
              TenantName = "CloudSphere"  },
            new Habit { Id = 102, Name = "Write every day",
              Description = "Finish your book project",
              TenantName = "CloudSphere"  }
        );
    }
}
```

The `GoodHabitsDbContext` class that was created previously had a single hardcoded database connection string. We will replace this and make use of the multiple database connections that we defined in the preceding config.

Replace the `GoodHabitsDbContext` class entirely with the following:

```
using Microsoft.EntityFrameworkCore;
using GoodHabits.Database.Entities;

namespace GoodHabits.Database;
public class GoodHabitsDbContext : DbContext
{
  private readonly ITenantService _tenantService;

  public GoodHabitsDbContext(DbContextOptions options,
    ITenantService service) : base(options) =>
    _tenantService = service;

  public string TenantName { get => _tenantService
    .GetTenant()?.TenantName ?? String.Empty; }

  public DbSet<Habit>? Habits { get; set; }

    protected override void
      OnConfiguring(DbContextOptionsBuilder optionsBuilder)
    {
        var tenantConnectionString =
          _tenantService.GetConnectionString();
        if (!string.IsNullOrEmpty(tenantConnectionString))
        {
            optionsBuilder.UseSqlServer(_tenantService
              .GetConnectionString());
        }
    }

    protected override void OnModelCreating(ModelBuilder
      modelBuilder)
  {
    base.OnModelCreating(modelBuilder);
    modelBuilder.Entity<Habit>().HasQueryFilter(a =>
      a.TenantName == TenantName);
    SeedData.Seed(modelBuilder);
    }

  public override async Task<int>
    SaveChangesAsync(CancellationToken cancellationToken =
    new CancellationToken())
  {
    ChangeTracker.Entries<IHasTenant>()
```

```
        .Where(entry => entry.State == EntityState.Added ||
           entry.State == EntityState.Modified)
        .ToList()
        .ForEach(entry => entry.Entity.TenantName =
           TenantName);
    return await base.SaveChangesAsync(cancellationToken);
    }
}
```

The main change in the preceding code is that the connection string is now read from the `TenantService` class that we created previously. This is far more dynamic and allows us to create new databases for new tenants on the fly as we build the app. This is also far more secure than having a connection string hardcoded in the source code and checked into the repository.

Another important change to note is that we add a query filter at the context level. This ensures that only the correct tenant can read their data, which is a very important security consideration.

Finally, we have overridden the `SaveChangesAsync` method. This allows us to set the tenant name here and not have to consider it in any of our other implementation code. This cleans up the rest of our code considerably.

Writing a service layer

We have now configured the `Habit` service and the database to work in a multi-tenant way, enforcing the presence of a tenant ID in every request. This is a good start to provide good security and separation between tenants.

Next, we'll demonstrate what we did by hooking up the database and the service and placing some test calls to the Habit service, showing how the tenancy is enforced.

We will start by writing a service layer. Open a terminal in the `HabitService` folder and run the following script:

```
touch IHabitService.cs; \
touch HabitService.cs;
```

Populate the interface with the following:

```
using GoodHabits.Database.Entities;
namespace GoodHabits.HabitService;
public interface IHabitService
{
        Task<Habit> Create(string name, string
           description);
        Task<Habit> GetById(int id);
        Task<IReadOnlyList<Habit>> GetAll();
}
```

Then, populate the class with the following:

```
using GoodHabits.Database;
using GoodHabits.Database.Entities;
using Microsoft.EntityFrameworkCore;

namespace GoodHabits.HabitService;

public class HabitService : IHabitService
{
    private readonly GoodHabitsDbContext _dbContext;
    public HabitService(GoodHabitsDbContext dbContext) =>
      _dbContext = dbContext;
    public async Task<Habit> Create(string name,
      string description)
    {
        var habit = _dbContext.Habits!.Add(new Habit { Name
          = name, Description = description }).Entity;
        await _dbContext.SaveChangesAsync();
        return habit;
    }

    public async Task<IReadOnlyList<Habit>> GetAll() =>
      await _dbContext.Habits!.ToListAsync();

    public async Task<Habit> GetById(int id) =>    await
      _dbContext.Habits.FindAsync(id);

}
```

This service is just a simple wrapper around some of the calls to the database. We could add a lot more functionality, but this will serve to demonstrate how multi-tenancy works in practice.

Writing the controller

With the service created, we'll now add a controller that makes the data from the service available over HTTP.

Run the following script in the HabitService folder to set up the required files:

```
rm WeatherForecast.cs; \
cd Controllers; \
rm WeatherForecastController.cs; \
touch HabitsController.cs; \
cd ..; \
```

```
mkdir Dtos; \
cd Dtos; \
touch CreateHabitDto.cs
```

Then, add the code for the controller, as shown here:

```
using GoodHabits.HabitService.Dtos;
using Microsoft.AspNetCore.Mvc;
namespace GoodHabits.HabitService.Controllers;
[ApiController]
[Route("api/[controller]")]
public class HabitsController : ControllerBase
{
    private readonly ILogger<HabitsController> _logger;
    private readonly IHabitService _habitService;
    public HabitsController(
        ILogger<HabitsController> logger,
        IHabitService goodHabitsService
        )
    {
        _logger = logger;
        _habitService = goodHabitsService;
    }
    [HttpGet("{id}")]
    public async Task<IActionResult> GetAsync(int id) =>
      Ok(await _habitService.GetById(id));
    [HttpGet]
    public async Task<IActionResult> GetAsync() => Ok(await
      _habitService.GetAll());
    [HttpPost]
    public async Task<IActionResult>
      CreateAsync(CreateHabitDto request) => Ok(await
      _habitService.Create(request.Name,
      request.Description));
}
```

This controller simply gives us two endpoints to create and read a habit from the database, via the service layer that we created previously.

Finally, add the following code to the `CreateHabitDto` file:

```
namespace GoodHabits.HabitService.Dtos;
public class CreateHabitDto {
    public string Name { get; set; } = default!;
    public string Description { get; set; } = default!;
}
```

Adding a service extension

Now that we are potentially dealing with many instances of the database, we need to add the ability to create and update all of the databases when the application starts up. We will create an extension to the service collection to facilitate this.

Add a class to the HabitService project called `ServiceCollectionExtensions`, and then add the following code:

```
using GoodHabits.Database;
using Microsoft.EntityFrameworkCore;

namespace GoodHabits.HabitService;
public static class ServiceCollectionExtensions
{
    public static IServiceCollection
      AddAndMigrateDatabases(this IServiceCollection
      services, IConfiguration config)
    {
        var options = services.GetOptions
          <TenantSettings>(nameof(TenantSettings));
        var defaultConnectionString =
          options.DefaultConnectionString;
        services.AddDbContext<GoodHabitsDbContext>(m =>
          m.UseSqlServer(e => e.MigrationsAssembly(
          typeof(GoodHabitsDbContext).Assembly.FullName)));

        var tenants = options.Tenants;
        foreach (var tenant in tenants)
        {
            string connectionString;
            if (string.IsNullOrEmpty(
              tenant.ConnectionString))
            {
                connectionString = defaultConnectionString;
            }
            else
            {
                connectionString = tenant.ConnectionString;
            }
            using var scope = services
              .BuildServiceProvider().CreateScope();
            var dbContext =
              scope.ServiceProvider.GetRequiredService<Good
              HabitsDbContext>();
            dbContext.Database.SetConnectionString(
```

```
        connectionString);
      if (dbContext.Database.GetMigrations()
        .Count() > 0)
      {
          dbContext.Database.Migrate();
      }
    }
    return services;
  }
  public static T GetOptions<T>(this IServiceCollection
    services, string sectionName) where T : new()
  {
      using var serviceProvider =
        services.BuildServiceProvider();
      var configuration =
        serviceProvider.GetRequiredService<
        IConfiguration>();
      var section = configuration.GetSection(
        sectionName);
      var options = new T();
      section.Bind(options);
      return options;
  }
}
```

The key point to understand from the preceding code is that the database connection string is set on a per-tenant basis, and that the tenants' database is updated per the latest migrations when that tenant logs in to the app.

This system takes much of the overhead of maintaining many databases away from the administrators. It is all handled automatically!

Application plumbing

Finally, we need to wire up all these new services. This is done in the `Program.cs` file in the `GoodHabits.HabitService` project. In this file, paste the following:

```
using GoodHabits.HabitService;
using GoodHabits.Database;
using GoodHabits.HabitService;
using Microsoft.OpenApi.Models;
var builder = WebApplication.CreateBuilder(args);
builder.Services.AddHttpContextAccessor();
builder.Services.AddControllers().AddNewtonsoftJson();
```

```
builder.Services.AddSwaggerGen(c => c.SwaggerDoc("v1", new OpenApiInfo
{ Title = "GoodHabits.HabitService", Version = "v1" }));
builder.Services.AddTransient<ITenantService, TenantService>();
builder.Services.AddTransient<IHabitService, HabitService>();
builder.Services.Configure<TenantSettings>(builder.Configuration.
GetSection(nameof(TenantSettings)));
builder.Services.AddAndMigrateDatabases(builder.Configuration);
builder.Services.AddEndpointsApiExplorer();
var app = builder.Build();
if (app.Environment.IsDevelopment())
{
    app.UseDeveloperExceptionPage();
    app.UseSwagger();
    app.UseSwaggerUI(c => c.SwaggerEndpoint(
       "/swagger/v1/swagger.json", "GoodHabits.HabitService
       v1"));
}
app.UseHttpsRedirection();
app.UseRouting();
app.UseAuthorization();
app.MapControllers();
app.Run();
```

In this code, you can see that we add the two new services that we created. We also make use of the service extension created previously and configure the Swagger endpoints.

Testing

With the previous configurations in place, we can now run the application and see how the changes we made have affected the operation of the controllers.

Because we have made some changes to the database (by adding the TenantName column to the Habit table), we will need to create a new migration with Entity Framework. Create the migration by navigating to the database project and running the following:

```
dotnet-ef migrations add MultiTenant --startup-project ../GoodHabits.
HabitService/GoodHabits.HabitService.csproj
```

Start the application with dotnet run and check that the API runs correctly.

You can look at the API in Swagger, but if you try to hit one of the endpoints, you will see an invalid tenant exception. This is expected, as we now have to add a header to every request to identify which tenant the request is for.

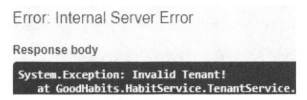

Figure 3.1 – The Swagger error

If all has gone well, the API will start, and you can view it in Swagger to see the preceding available endpoints!

In order to test HabitService, we will use the ThunderClient extension that we installed in *Chapter 2*.

Open the ThunderClient extension, click on **New Request**, and set the request up as shown, with the tenant specified as CloudSphere.

Figure 3.2 – The ThunderClient request

The preceding is a properly configured request with the tenant added. Hitting **Send** will issue the response, and if everything has worked, you should see the following response:

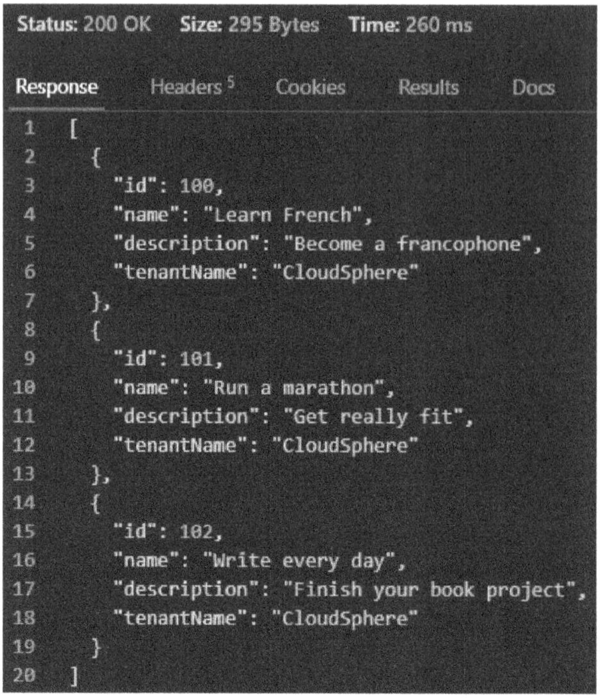

```
Status: 200 OK    Size: 295 Bytes    Time: 260 ms

Response       Headers 5    Cookies     Results     Docs
 1   [
 2       {
 3         "id": 100,
 4         "name": "Learn French",
 5         "description": "Become a francophone",
 6         "tenantName": "CloudSphere"
 7       },
 8       {
 9         "id": 101,
10         "name": "Run a marathon",
11         "description": "Get really fit",
12         "tenantName": "CloudSphere"
13       },
14       {
15         "id": 102,
16         "name": "Write every day",
17         "description": "Finish your book project",
18         "tenantName": "CloudSphere"
19       }
20   ]
```

Figure 3.3 – The ThunderClient response

The preceding shows that we have successfully returned the data for the `CloudSphere` tenant.

You should take some time now to do some more investigations with this and convince yourself that we have correctly limited access to data on a per-tenant basis!

That is the end of the practical part of this chapter. We'll now move on to consider the security implications of what we have done.

Security considerations

I think that it should go without saying that multi-tenancy significantly affects the security landscape of a SaaS application.

There are many aspects to security, ranging from simple (or at least standard) considerations, such as authenticating users to the ability of the application, to withstanding very rare events, such as a natural disaster taking out a crucial data center or a vital internet backbone.

The word "security" in this context refers to the overall ability of an application to withstand anything that the world may throw at it – and continue to keep the tenants' and users' data correct, accessible, and private! This can include hacks, attacks, natural disasters, coding errors leaking data, or even regulatory issues, such as the **General Data Protection Regulation (GDPR)**.

Potential security pitfalls

Developers of every type of application will face a broad range of potential security issues. It is an unfortunate reality of SaaS applications and, particularly, multi-tenant applications that the developers of such will have to consider nearly all possible security threats!

SaaS applications are typically layered, with at least a database, an API, and a user interface. Each of these layers presents an attack vector. SaaS applications are also often hosted on a cloud service. Hosting in the cloud is generally more secure than hosting on-premises, but there are a number of additional attack vectors that must be considered. Most importantly, the user accounts of the administrators can be compromised and the admin pages accessed remotely. This is generally less of a concern with on-prem solutions, which may have no remote access at all.

This section will list some of the security considerations that you will have to take into account as a developer of SaaS applications.

Resource access management

First, and most obviously, access to the resources in a system must be limited to only those with permission to see those resources.

For example, and rather obviously, data pertaining to one tenant should never be visible to a user on another tenant, unless there is a case where data is deliberately shared. When developing a SaaS application, it's important to understand that an attacker will target the user interface, the API, and also the database in an effort to gain access to the data.

An attacker will also attempt to intercept data when it is "in flight." When data moves from the database to the API, or from the API to the user interface, it is vulnerable.

As well as individual user data, it is likely that there will be sections of an application that are only accessible by users in a certain *role*, such as administrators.

Any SaaS system that fails to control resource access will very quickly be compromised, and an application with a reputation for losing customers' data will quickly cease to attract customers. Failures in this area can be catastrophic to an application and, indeed, a company's reputation.

This is a very important consideration in a SaaS application, particularly a multi-tenant application where tenants will share access to certain resources (such as the database), and access to these resources must be managed very carefully.

Data leaks

Similar to the aforementioned, data leaks can prove very costly in terms of reputation, and they can also have hugely significant financial implications for the business that owns the application!

While a data leak is, in some ways, a failure of resource access, it is normally a more general problem. Some examples to consider are as follows:

- A user interface that transmits credit card details from the client to the server in plain text, allowing a **man-in-the-middle** (**MitM**) attack that gathers the credit card information from all users. As stated previously, data "in flight" tends to be vulnerable.

- An API endpoint that misses the authentication attribute, thus rendering it accessible by anyone, whether they are authenticated or not. Similarly, an API endpoint may have an incorrect authentication attribute. This is extremely important, and measures should be put in place to automatically test that the API is appropriately secured.

- A database with poor security that allows an attacker to access, download, or delete the data in the database.

- A server or a VM that has been improperly configured and can be accessed remotely by a malicious user.

- Poor password hygiene that allows attackers to *guess* the passwords of users, either by brute force, rainbow tables, or similar *blunt-force* attacks.

Very often these days, data breaches can come with a very large financial penalty for a company that has operated an insecure application. In 2018, the **British Airways** (**BA**) website leaked the credit card data of 380,000 customers – one of whom would go on to write this book! The financial cost to BA was £183 million ($244 million at the time of writing the book). It would have been considerably cheaper for BA to hire a team of developers and security specialists to secure its website in the first place!

There is an additional consideration when building a SaaS application, in particular an application with a multi-tenant database. A large risk vector for these types of applications in terms of data leaks is that one tenant's data will accidentally be revealed to a user belonging to a different tenant when those two tenants share a storage resource, an app server, a database, or similar. As discussed in this chapter, a rigorous approach to designing a multi-tenant solution when the app is in the concept stage is needed!

Data corruptions

The various users and tenants in a multi-tenant application are separated from each other in one of the ways described previously, but the reality is that they are often sharing some (or many) of the same resources. They may be sharing a database, an app server, a user authentication system, and so on. Should one of these systems be corrupted by one user, there may be some contagion, and the corruption may spread and subsequently impact another user. This is rare in the modern era where cloud resources are typically used, and the big cloud providers should have a handle on these types of issues.

The primary mitigation for this is to be very careful when segregating individual tenant's concerns and to use a reputable cloud provider.

It goes without saying that you should back up your data regularly and also check that an application can be recovered from said backups!

Hacks and attacks

While some of the aforementioned data leaks could be classed as hacks or attacks, a data leak is typically more *passive* – like a pipe that is leaking water. A hack is more active – like a pipe that someone has hit with an axe!

There is a huge range of attacks that the operators of SaaS applications may face, but deep-diving security concerns are out of the scope of this chapter and, indeed, this book.

However, it is worth considering the possibility of a co-tenant attack, as they are specific to a multi-tenant application. While an external attacker will typically have some access to the application, assuming that, as a minimum, the login page is available over the public internet, a fully paid-up user will by definition have more access to more parts of the system; therefore, they will be in a better position to mount an attack. This attack could be against the infrastructure or other users.

Configuration

SaaS applications add a degree of complexity, and with this complexity will come a certain amount of configuration overhead.

A configuration error could, for example, accidentally leak a production password or access to a database or other resource.

It is very easy to overlook configuration parameters, but this can be a costly mistake. Everything should be stored in as secure a way as possible, making use of key vaults and secret managers as much as possible.

Care should also be taken when it comes to checking configuration files in the source code repository, as this is unfortunately a common way to leak passwords and the like.

Storage

Every SaaS application will have some form of data stored, and that data presents an attack vector that can be utilized to compromise the application.

Data retention and deletion

Similar to the aforementioned, a SaaS application will retain sensitive data. It is vitally important that this is secured. Another key consideration with data retention is that in many jurisdictions, individuals have the "right to be forgotten," where all of the data relating to that individual must be anonymized.

This can be very challenging in the context of a SaaS application for the following reasons:

- The database will typically be multi-tenant. How easy is it to completely wipe out all of the information relating to one tenant without orphaning other records?

- Logging will most likely be performed across all layers – user interface, API, and database. How sure can you be that there is no sensitive customer data in those logs?

- Backups are obviously a very important part of the security of a system, but should you have to do a restoration at any point, how can you be sure that previously deleted data has not now been restored?

All of the aforementioned require a meticulous approach to data retention and deletion in order to guarantee your customers that their data is secure.

Regulations

A huge part of data security when building a modern SaaS application is to understand the regulatory frameworks that you must comply with. The internet offers a truly global reach for your application, but that global reach means that you may have to consider a huge array of regulatory frameworks, of which the consequences for breaching are often astronomically high fines and penalties.

The European Union's GDPR law stipulates rules for how organizations must use personal data, and the penalties for breaching these regulations can be incredibly harsh – 10 million euros, or 2% of a firm's entire annual global turnover. These kinds of laws are becoming increasingly common in many jurisdictions.

Key questions to ask yourself in this regard are as follows:

- What jurisdictions govern the data that will be stored in this application?

- Do your applications adhere to these requirements?

- Do the partners and third parties that you are working with adhere to these regulations?

Good security practices for multi-tenant applications

I hope that the preceding set of potential horror stories hasn't put you off! In this section, I'll list some of the mitigations that should be considered.

Governance, risk, and compliance (GRC) protocols

GRC is a set of processes and procedures that a business will put in place for many reasons, such as for meeting objectives, maintaining regulatory compliance, or adequately securing a SaaS application! These could be as follows:

- Procedures that detail how an application upgrade should be performed and a list of pre-release checks.

- Procedures that manage and control access to your customers' private data or IPs.

- Restrictions on which members of staff have access to production databases. Note that, ideally, no one should!

Isolation of assets and resources

As much as is practically possible, resources should be isolated from each other. Obviously, the app layer will need some access to the database; otherwise, how will it access data? However, it is good practice to run the database and the application on different servers or VMs. Note that isolation can bring in performance issues if not done correctly – this is a delicate balance!

With multi-tenant applications, individual users' and tenants' data should also be isolated as much as possible from other users and tenants. Ideally, all user data should be encrypted such that only the owner of the data can make sense of it. In this way, even if a data leak does occur, it's not possible to infer anything of value from the data leak.

Auditing

This is good practice for all applications, not just SaaS/multi-tenant. Independent third parties should be consulted to audit the security and compliance of all IT systems and also your tenants' data.

As well as checking that data is physically secure, auditors should also confirm that industry standards are being followed and that any regulatory frameworks are adhered to.

Using Data Loss Prevention (DLP) software

There are many commercial DLP solutions available. It is worth considering delegating this complex and extremely important to a specialist provider.

Managing access carefully

A SaaS/multi-tenant application is particularly susceptible to security issues arising from improperly allocated access to a resource. This is true from a company point of view – developers should not have access to production databases. This is also true from a user point of view.

User access should be managed based on identities (who are the individual users?) and also through roles (what types of user classes do the individuals fall into?).

Being very careful with collaborative tools

Often, a SaaS application will allow users to choose to share certain resources with other selected users within the same tenant, and less commonly with users in a different tenant. This can be an incredibly effective addition to a SaaS app and is indeed a massive advantage to using a multi-tenant solution. However, this introduces a security risk whereby the application itself can programmatically leak data. Significant additional care should be taken when building collaborative elements in a SaaS app, as these will be the first place a bad actor will look to exploit access to another user's data.

Good security is hard in any application, and doubly so when dealing with multi-tenant applications. As ever, it is much easier to start a project with good practices in place than it is to add them later!

Summary

As we can see, multi-tenancy is a large and complex topic but one that is absolutely core to understanding when building a SaaS application.

The decisions that are made on how a database is segregated will have the largest impact on the rest of an application, but it is very important to consider how the presence of multiple tenants will affect the API and the user interface layers as well.

Multi-tenancy introduces considerable additional security considerations. These should be thought of upfront and kept in mind throughout the development process.

Having said this, the additional challenges of implementing a multi-tenant solution provide *massive* potential upsides! The reach of an application can be truly global, and the scalability is unmatched by any other paradigm. Consider the biggest tech companies in the world – Google, Meta, Netflix, and so on. Every one of these companies has embraced the idea of SaaS and multi-tenancy, and they have done so for a good reason!

In the next chapter, we will build on what we have learned about multi-tenancy and learn about how to build databases and plan for data-rich applications.

Further reading

- Single Tenant vs Multi Tenant: SaaS Architecture: `https://www.clickittech.com/aws/single-tenant-multi-tenant/`

- Multi-tenancy strategies: `https://www.linkedin.com/pulse/effective-multi-tenancy-strategies-saas-applications-kulkarni/`

- Strategies to build a scalable multi-tenant SaaS solution: `https://aws.amazon.com/blogs/apn/in-depth-strategies-for-building-a-scalable-multi-tenant-saas-solution-with-amazon-redshift/`

- Implementing multi-tenancy SaaS apps: `https://developers.redhat.com/articles/2022/05/09/approaches-implementing-multi-tenancy-saas-applications`

- A definition of multi-tenancy: `https://www.techtarget.com/whatis/definition/multi-tenancy`

- How multi-tenancy affects embedded analysis: `https://yurbi.com/blog/what-is-multi-tenancy-security-and-how-does-it-impact-embedded-analytics/`

- DLP definition: `https://digitalguardian.com/blog/what-data-loss-prevention-dlp-definition-data-loss-prevention`

4
Building Databases and Planning for Data-Rich Applications

In previous chapters, we established a foundational SaaS application consisting of a straightforward database with a single table. This database was connected to an API, and we showcased secure multi-tenancy implementation using SQL Server, .NET, and Entity Framework.

In this chapter, we will delve deeper into the intricacies of the database layer and its interactions with Entity Framework. As the bedrock of an entire application, the design choices made at the database level will influence every subsequent layer in the stack. We will explore how to construct and design resilient databases for data-intensive SaaS applications. You will acquire a variety of skills, such as normalization, indexing, performance optimization, as well as techniques to test and maintain the database.

Once deployed, a database often represents the most demanding aspect of a system to keep current. The database is inherently stateful, and it is crucial to prevent data loss or corruption during updates. In addition to learning about database design and construction, we will examine several strategies to maintain and update the database, with an emphasis on the tools provided by Entity Framework.

The expertise you gain in this chapter is vital to create scalable and dependable SaaS applications. By mastering the methods discussed in this chapter, you will be capable of designing databases optimized for performance, scalability, and maintainability, thereby facilitating the development and maintenance of your SaaS applications.

This chapter covers the following main topics:

- The importance of data in a SaaS application

- Building a database using SQL Server and Entity Framework

- Testing the database and data-rich applications

- Working in production, keeping your database up to date, and keeping your data safe

Data and a database are the foundations of a SaaS application. Let's start by considering how important they are.

Technical requirements

All code from this chapter can be found at `https://github.com/PacktPublishing/Building-Modern-SaaS-Applications-with-C-and-.NET/tree/main/Chapter-4`.

The importance of data in a SaaS application

In a SaaS application, a database serves as the foundation for the application. The database (or more accurately, the data contained within) is what drives the application and where the primary value for users lies. A SaaS application without data is just an empty shell!

One of the key considerations when building a SaaS application is the type of data that the application will store and how it will be used. Will the application be storing large amounts of structured data, such as customer records or transaction histories? Or will it be storing unstructured data, such as the data underpinning a social media feed? The type of data will have a significant impact on the design and architecture of the application.

Another important consideration is how the data will be accessed and manipulated. Will the data be accessed by a large number of users simultaneously, or will it only be accessed by a few users at a time? Will the data be updated frequently, or will it be mostly static? These factors will influence the choice of database technology and the design of the data model.

In a SaaS application, it is also important to consider how the data will be shared among different tenants. As we discussed in the previous chapter, the data for each tenant must be kept separate and secure, while still allowing for efficient access to and manipulation of the data that pertains to the current tenant. This requires careful planning and design of a data model and database schema.

It is also important to consider scalability. While a SaaS application may well start off with a small user base and thus a comparatively low number of reads/writes, this can change very quickly as the user base increases! It's important to design a data model and schema in a way that allows the application to grow. Similarly, the amount of data that is transmitted over the internet must be managed. Bandwidth is not free nor unlimited, and in data-intensive applications, this can become a burden.

The importance of data is not limited to technical considerations. Data plays a crucial role in the user experience. The way that data is presented, organized, and accessed can significantly impact the usability of an application. While this is, of course, dependent on the user interface, the structure of the underlying data, and the ease and speed with which it can be queried, will be noticed at the frontend by end users.

For a company building an application and hosting a database, data can be a key source of revenue. Some SaaS applications monetize their data by selling access to information about consumer demographics and buying habits to businesses looking to target specific market demographics. This data is often collected and classified by machine learning algorithms, allowing for insights about the users and creators of the data. With this valuable information, businesses can create targeted marketing campaigns and improve their products and services to better meet the needs of their customers.

Data is important for a myriad of reasons, and so it should go without saying that maintaining data security and compliance is a crucial aspect of building a successful SaaS application. It is the responsibility of the SaaS provider to ensure that sensitive data, such as financial or personal information, is kept secure and compliant with relevant regulations. To achieve this, the SaaS provider may need to implement various security measures, such as encryption and access controls, to protect the data.

Data and the database are absolutely critical parts of a SaaS application, and it's important for SaaS developers to have a strong understanding of how to design, build, and maintain data-rich applications.

Building a database

In this section, we will focus on using SQL Server and Entity Framework to design and construct a database for your SaaS application. We will cover a range of topics, including choosing the right database technology, designing an efficient and scalable data model with Entity Framework, and implementing database security and compliance measures with SQL Server. By the end of this section, you will have a solid understanding of how to build a robust and reliable database for your SaaS application.

Types of database

Because this is a book focused on making use of the Microsoft stack, I will focus on SQL Server as the underlying database platform, and I will use Entity Framework to interact with the database. SQL Server is a **relational database**, which is a type of database that stores data in the form of tables, with rows representing individual records and columns representing data attributes. It is a very structured way to store data, and the "shape" of the data must be known in advance and built into the application at design time.

While we will focus on SQL Server and, therefore, relational data, it is worth briefly considering the alternatives, which are outside the scope of this book. Some of the following alternatives may be worth further investigation if you have a specific use case that may require something other than relational data:

- **Document databases**: A document database is a type of database that stores data in the form of documents. In this context, a document is data that is typically made up of key-value pairs and is designed to be scalable and flexible. In a document database, the structure or shape of the data will not be set when the database is designed, making it a good choice to store and query large volumes of data with diverse structures.

- **Graph databases**: These databases store data in the form of nodes (representing data entities) and edges (representing relationships between nodes). They are often used in applications that need to represent complex relationships between data entities, such as social networks or recommendation engines.

- **Key-value stores**: These databases store data in the form of key-value pairs, where the key is used to identify the data and the value is the data itself. They are often used for simple data storage and retrieval scenarios where the data does not need to be queried or indexed. This is similar to a document database but more limited in that it is only really suitable for simple use cases.

- **In-memory databases**: These databases store data in memory, rather than on disk. They are often used for applications that require fast read and write access to data, such as online gaming or financial applications. Note that in-memory databases can also be used to help test databases. These are two separate use cases and should not be confused.

- **Time-series databases**: These databases are designed specifically for storing and querying time-stamped data, such as sensor readings or financial transactions. They are often used in applications that need to analyze data over time.

In this chapter, we will focus on SQL Server, which is a relational database. We will interact with the database using a tool that Microsoft developed specifically for this purpose – namely, Entity Framework.

What is ACID?

When working with databases – and particularly relational databases – you will often come across the acronym **ACID**. This refers to the four properties of a database transaction – namely, atomicity, consistency, isolation, and durability:

- **Atomicity**: A transaction is treated as a single, indivisible unit of work, and either all its operations are completed or none of them are

- **Consistency**: The transaction brings the database from one valid state to another, preserving database invariants and constraints

- **Isolation**: The concurrent execution of transactions results in a system state that would be obtained if transactions were executed serially, in some order

- **Durability**: Once a transaction has been committed, its changes to the database persist and survive future system failures

These properties are a set of properties that are guaranteed by a database management system to ensure the reliability and consistency of data stored in a database. ACID is most commonly associated with **relational database management systems** (**RDBMS**), such as Oracle, MySQL, PostgreSQL, and Microsoft SQL Server. However, some newer databases, such as NoSQL databases and NewSQL databases, may also provide ACID guarantees, although they may have different levels of consistency and durability. The level of ACID support depends on the specific database technology and how it is implemented and configured.

ACID is generally associated with transactions in relational databases, and less commonly so in NoSQL or document databases. In this chapter, and indeed throughout this book, we will focus on SQL Server, a relational database that provides support for ACID transactions.

Entity Framework

Entity Framework is an **object-relational mapping** (**ORM**) tool that allows developers to interact with a database using .NET objects. It simplifies the process of accessing and manipulating data by eliminating the need to write SQL queries and manually map data to objects. Entity Framework is well-suited for developers who are familiar with .NET and want to streamline their data access and manipulation tasks, which makes it an excellent choice for study in this .NET-focused book!

While we will focus on SQL Server, one of the big benefits of using Entity Framework is its ability to generate database-agnostic code, allowing developers to change a database platform or support multiple database platforms without having to significantly rewrite their code. This feature is of particular interest when dealing with a multi-tenant SaaS application, where certain customers may mandate a specific database platform.

Entity Framework wraps up a lot of the complexities associated with writing code that interacts with a database. Concepts such as lazy loading, change tracking, and automatic migration of data and schema changes are handled out of the box.

Unlike many other ORMs, Entity Framework supports several different approaches to interacting with a database, including the traditional database-first approach, the code-first approach, and the model-first approach. This gives developers the flexibility to choose the approach that best fits their needs.

Entity Framework is a powerful tool that can greatly enhance the productivity of experienced .NET developers by simplifying data access and manipulation tasks, and it is highly recommended when approaching a project that will be highly dependent on a data platform – such as a SaaS application.

It is out of the scope of this chapter to cover all of the possible ways to use Entity Framework, so I will focus on one – Code First.

Code first with Entity Framework

Code first with Entity Framework is a development approach that allows developers to create their .NET application's data model using C# code, rather than designing a database using a UI such as SQL Server Management Studio, or through writing and maintaining SQL scripts. This approach is particularly useful for developers who prefer to work with code and want more control over their application's data model. With code first, developers can define their data model using classes and properties in their code, and Entity Framework will handle the creation and management of the underlying database. This approach allows developers to focus on the data model and business logic of their application, without having to worry about the implementation details of the database.

We saw this approach in *Chapter 2*. When the database was configured, we didn't write a single line of SQL code – we wrote a C# class called `GoodHabits.cs` and defined the data structure in C# code. We then used two commands on the Entity Framework CLI that updated the database. In *Chapter 3*, we modified this file to facilitate multi-tenancy.

Stored procedures with Entity Framework

Traditionally, it has been very common to use stored procedures when designing a database. While this is still a very valid and useful approach to database development, it is increasingly being seen as best practice to use an ORM such as Entity Framework to access and manipulate data in a database, rather than using stored procedures. There are a few reasons for this.

Entity Framework allows you to work with objects and entities in your code, rather than having to write raw SQL queries. This gives you a higher level of abstraction, which can make it easier to develop an application. With this approach, you can build your database in a familiar object-orientated way, which can make it easier to reason about and maintain. Entity Framework is able to interpret relationships between objects and create database relationships from C# code. If you create logic and then model it in a stored procedure, Entity Framework loses sight of that logic.

Another huge benefit of using Entity Framework is that many database platforms are supported out of the box. However, logic in stored procedures is typically not transferrable between database platforms and would have to be built and maintained separately on a per-platform basis.

Finally, Entity Framework has a number of testing tools available to use. Using stored procedures and triggers will require specific testing infrastructure and techniques, and this may make it harder to test an application because logic is split between code and the database.

There are certain cases when using stored procedures or triggers may be beneficial. These include the following:

- When working with very large datasets, or in situations where performance is critical, it may be advantageous to execute logic directly against a database by way of a stored procedure
- In cases where data security is a concern, stored procedures can help to prevent unauthorized access to data by limiting the types of queries that can be run against a database.

- In cases where you want to abstract the data access layer of your application from the underlying database schema, using stored procedures can help to decouple the two. This can be particularly useful in situations where a database schema may change frequently.

It is important to understand your specific use case when deciding whether or not to make use of stored procedures in your database. For the demo application, we will continue to use Entity Framework to manipulate and access the database.

Normalization

Database normalization is the process of organizing a database in a way that reduces redundancy and improves data integrity. It involves dividing the database into smaller, more focused tables that are related to each other through relationships. The goal of normalization is to eliminate redundancy and ensure that each piece of data is stored in only one place in the database. Normalization is an important step in the database design process and can greatly improve the performance and reliability of a database.

Entity Framework supports the process of normalizing a database in several ways. One of the main ways it does this is through the creation and modification of tables and relationships within the database. This allows developers to structure their data in a way that reduces redundancy and improves data integrity – a key goal of normalization. Entity Framework also includes support for the automatic migration of data changes. This means that when developers make changes to their data model, those changes are automatically reflected in the underlying database. This can be particularly useful when normalizing a database, as it allows developers to easily update the structure of their data without having to manually migrate data between tables.

In addition, Entity Framework's LINQ query syntax allows developers to easily retrieve and manipulate data from a normalized database. It supports a wide range of operations, including filtering, sorting, and aggregation, making it easy to work with data from multiple tables. Finally, Entity Framework's support for eager and lazy loading allows developers to optimize the performance of their application by only loading the data they need on demand, rather than loading all data upfront. This can be harder with a poorly normalized database. Overall, Entity Framework provides a number of tools and features to help developers normalize their databases and improve the performance and reliability of their applications.

There are several normal forms that can be used to measure the degree of normalization in a database. The first three (called 1NF, 2NF, and 3NF) are used to address redundancy in a database and are generally considered to be good practice in most instances.

Beyond the first three, the additional normal forms are designed to address specific types of problems; however, these are less commonly used and are considered out of the scope of this chapter.

It is worth noting that achieving higher normal forms does not always constitute a better-designed database. It is generally better to design a database that is efficient and performant around a specific use case than to adhere blindly to the normalization rules. That said, achieving 3NF is usually a good starting point from which to work, and further normalization, or indeed denormalization, can follow from there.

Let's illustrate this with an example. Let's consider adding a user table to the GoodHabit example we developed in *Chapter 2*.

To achieve the first normal form (1NF), all attributes in the database must be atomic. This means that each column in the database should contain a single value. We could design a user table that looked like this:

ID	Name	Habits
1	Roger Waters	Learn French [Daily], Go Running [Weekly]
2	Dave Gilmore	Play Guitar [Daily], Learn French [Weekly]
3	Nick Mason	Play Drums [Daily]

Figure 4.1 – Poorly normalized data

The preceding table shows poorly normalized data. The **Name** column contains two pieces of information (first and last name), which may be useful to use separately. The **Habits** column contains a comma-separated list of data. This can be improved upon like so:

ID	First Name	Last Name	Habit	Frequency
1	Roger	Water	Learn French	Daily
2	Dave	Gilmore	Play Guitar	Weekly
3	Nick	Mason	Play Drums	Daily
4	Roger	Water	Go Running	Weekly
5	Dave	Gilmore	Learn French	Daily

Figure 4.2 – Data in the first normal form (1NF)

The preceding table shows the data in 1NF. Each attribute holds a single value. We now have a row in the database for each habit, but the users appear multiple times. If Dave decided that he would prefer to be called David, we would have to update the data in multiple places.

To move this data into the second normal form, we need to break the data into two tables – one for the user, and one for the habit. We will need a third table to link the users to the habits they are going to work on:

User		
ID	First Name	Last Name
1	Roger	Water
2	Dave	Gilmore
3	Nick	Mason

Habit		
ID	Habit	Frequency
1	Learn French	Daily
2	Play Guitar	Weekly
3	Play Drums	Daily
4	Go Running	Weekly

UserHabit		
ID	User ID	Habit ID
1	1	1
2	1	4
3	2	2
4	2	1
5	3	3

Figure 4.3 – Data in the second normal form (2NF)

This is much better, and we can start to see that this could be queried and updated in a very tidy way. There is one further improvement we could make though. The **Habit** table has a **Frequency** column that is indirectly dependent on the ID column. This is called a transitive dependency. In order to move this data to the third normal form, we must break this transitive dependency by adding a **Frequency** table:

User		
ID	First Name	Last Name
1	Roger	Water
2	Dave	Gilmore
3	Nick	Mason

Habit	
ID	Habit
1	Learn French
2	Play Guitar
3	Play Drums
4	Go Running

UserHabit			
ID	User ID	Habit ID	Frequency ID
1	1	1	1
2	1	4	2
3	2	2	1
4	2	1	2
5	3	3	1

Frequency	
ID	First Name
1	Daily
2	Weekly

Figure 4.4 – Data in the third normal form (3NF)

The third normal form is sufficient at this stage, and we'll take this no further. You can see that all of the data is separated into individual tables, and linked through foreign key constraints to the **UserHabit** table. This allows efficient querying and updating of the data in the database.

Indexing and performance optimization

Indexing a database means creating a separate data structure that is used to improve the performance of certain types of queries. Indexes are typically created on specific columns within a table, allowing the database to quickly locate the rows that match given criteria. For example, if you have a large table of customer records and you frequently search for customers by their last name, you could create an index on the `last_name` column to improve the performance of those searches. Indexes can significantly improve the performance of certain types of queries, but they also have a cost in terms of storage space and maintenance. As a result, it is important to carefully consider which columns should be indexed and to balance the benefits of indexing with the costs.

To do indexing with Entity Framework, developers can use a variety of tools and approaches. One way to do indexing is to use the Entity Framework Fluent API, which allows developers to define indexes on their entities using code. To create an index using the Fluent API, developers can use the `HasIndex` method and specify the properties that should be included in the index. Another option is to use Entity Framework Designer, a visual tool that allows developers to design their data model using a graphical interface. The Designer includes the ability to define indexes on entities by right-clicking on an entity and selecting **Add Index**. Finally, developers can use database migrations to create indexes on their entities by adding the appropriate code to their migration files.

Configuring indexes with Entity Framework is straightforward!

If we consider the GoodHabit table that we developed in *Chapter 2*, we used the following C# code to define the structure of the table:

```
public class GoodHabit
{
  public int Id { get; set; }
  public string Name { get; set; } = default!;
}
```

We can add an index to the Name column by decorating the class with an attribute like so:

```
[Index(nameof(Name))]
public class GoodHabit
{
  public int Id { get; set; }
  public string Name { get; set; } = default!;
}
```

This will instruct the database platform to create an index for the Name column. We could do the same for the ID column in the same way. You can create a composite index by setting the attribute as follows:

```
[Index(nameof(Name), nameof(Id))]
```

If you need to set the sort order of the indexes, you can use one of the following:

```
[Index(nameof(Name), nameof(Id)), AllDescending = true]
[Index(nameof(Name), nameof(Id)), IsDescending = new[] { false, true
}]
```

If you want to name your index, you can use the following:

```
[Index(nameof(Name), Name = "Index_Name")]
```

There is a great deal of flexibility provided by Entity Framework, and it is out of the scope of this book to cover all of it.

You can get more information about it at https://learn.microsoft.com/en-us/ef/core/modeling/indexes?tabs=data-annotations.

We will now build out a database for the example application, being mindful of what we have just learned.

Designing a database schema using Entity Framework code first

Before we start designing a database, let's stop and think about what our requirements are:

- The database will store user information to identify individual users
- Users can add habits to the database to track their progress and achieve their goals

- Users can log progress for each habit and set reminders to perform it

- Users can set goals for their habits, such as running a certain number of miles or saving a certain amount of money

- The database can support users in achieving their goals and improving their overall well-being by tracking and managing their habits

The following figure shows a diagram representing the database that we just configured. This shows the tables, columns, and relationships between the tables:

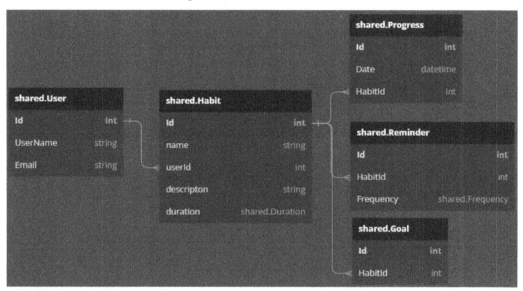

Figure 4.5 – A suggested database schema

The preceding diagram shows a schema that can meet the requirements defined previously.

Creating the entity classes

As we are using Entity Framework code first, we will build this database by writing C# code. Create the entity classes with the following:

```
cd GoodHabits.Database/Entities; \
touch Goal.cs; \
touch Progress.cs; \
touch Reminder.cs; \
touch User.cs;
```

We already have the `Habit` entity, but we will update it with some additional properties. Copy the following code into `Habit.cs`:

```
using Microsoft.EntityFrameworkCore;
namespace GoodHabits.Database.Entities;

[Index(nameof(Name), nameof(Id))]
public class Habit : IHasTenant
{
    public int Id { get; set; }
    public string Name { get; set; } = default!;
    public string Description { get; set; } = default!;
    public int UserId { get; set; }
    public virtual ICollection<Progress> ProgressUpdates {
      get; set; } = default!;
    public virtual ICollection<Reminder> Reminders { get;
      set; } = default!;
    public virtual Goal Goal { get; set; } = default!;
    public Duration Duration { get; set; }
    public string TenantName { get; set; } = default!;
}

public enum Duration { DayLong, WeekLong, MonthLong }
```

Next, `Goal.cs` should look like this:

```
using Microsoft.EntityFrameworkCore;

namespace GoodHabits.Database.Entities;

[Index(nameof(Id))]
public class Goal
{
    public int Id { get; set; }
    public int HabitId { get; set; }
    public virtual Habit Habit { get; set; } = default!;
}
```

`Progress.cs` should look like this:

```
using Microsoft.EntityFrameworkCore;

namespace GoodHabits.Database.Entities;
[Index(nameof(Id))]
public class Progress
{
```

```
    public int Id { get; set; }
    public DateTime Date { get; set; }
    public int HabitId { get; set; }
    public virtual Habit Habit { get; set; } = default!;
}
```

Reminder.cs should look like this:

```
using Microsoft.EntityFrameworkCore;
namespace GoodHabits.Database.Entities;

[Index(nameof(Id))]
public class Reminder
{
    public int Id { get; set; }
    public Frequency Frequency { get; set; }
    public int HabitId { get; set; }
    public virtual Habit Habit { get; set; } = default!;
}
public enum Frequency { Daily, Weekly, Monthly }
```

And finally, User.cs should look like this:

```
using Microsoft.EntityFrameworkCore;
namespace GoodHabits.Database.Entities;
[Index(nameof(Id))]
public class User
{
    public int Id { get; set; }
    public string FirstName { get; set; } = default!;
    public string LastName { get; set; } = default!;
    public string Email { get; set; } = default!;
}
```

Add the following lines of code into the GoodHabitsDbContext class:

```
    public DbSet<User>? Users { get; set; }
    public DbSet<Progress>? Progress { get; set; }
    public DbSet<Reminder>? Reminders { get; set; }
    public DbSet<Goal>? Goals { get; set; }
```

You can create the migration by running the following in the console:

```
dotnet-ef migrations add AdditionalEntities --startup-project ../
GoodHabits.HabitService/GoodHabits.HabitService.csproj
```

The migrations will be automatically applied to the database when `HabitService` runs, so simply run the Habit service to push the changes to the database.

Using the Server Explorer to view the database, we can see that the schema has been successfully applied!

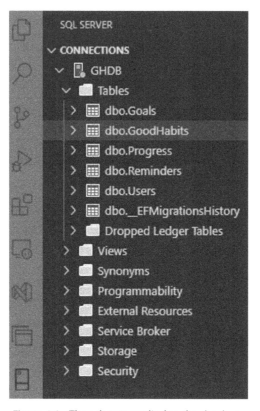

Figure 4.6 – The schema applied to the database

The preceding figure shows that the schema has successfully migrated to the database.

Testing data-rich applications

Testing the database layer of a SaaS application is an essential part of the development process. A database is a critical component of any application, as it stores and manages the data that an application relies on. Ensuring that the database is working correctly is crucial for the overall stability and performance of the application.

There are several challenges that you may encounter when testing the database layer of your application. One challenge is ensuring that the database schema is correct and that the data is stored and retrieved correctly. Another challenge is ensuring that the database is properly optimized for performance, particularly if you are dealing with large amounts of data.

There are a number of different techniques that you can use to test the database layer of your application. One common technique is to use unit tests to verify that individual database functions are working correctly. Another technique is to use integration tests to ensure that the database is working correctly in conjunction with the rest of the application. You may also want to use performance tests to ensure that the database is able to handle large amounts of data without experiencing any issues.

Unit tests are a type of automated testing that is used to verify the behavior of individual units of code, such as individual functions or methods. In contrast to unit tests, integration tests focus on testing how different parts of an application work together as a system. Integration tests are used to ensure that different components of the application are able to communicate and interact with one another correctly.

Upgrading a database

Upgrading a database with Entity Framework involves making changes to the database schema and data to reflect changes in an application. This can include adding new tables or columns, modifying existing tables or columns, and migrating data from one format to another.

There are several approaches that you can take when upgrading a database with Entity Framework. One approach is to use the DbMigrations class to automatically generate and execute the necessary SQL commands to update the database schema. This can be convenient, as it allows you to make changes to the database using a high-level API, rather than having to write raw SQL commands. However, it can also be less flexible than some other approaches, as it relies on Entity Framework to generate the SQL commands, and that may not always produce optimal results. This is the approach that we will take in this chapter.

It is worth being aware of a popular alternative approach, which is to use Entity Framework's DbContext class to manually execute SQL commands to update a database. This can be more flexible, as you have complete control over the SQL commands that are executed. However, it can also be more time-consuming and error-prone, as you have to write and debug the SQL commands yourself.

Finally, it is, of course, possible to update a database separately from Entity Framework using any preferred method, which most likely involves executing SQL scripts.

I think that it is generally preferable to make use of the built-in migration tools and allow Entity Framework to do the heavy lifting for us.

We saw this process in practice twice – when we created the initial migration in *Chapter 2*, and when we just updated the schema.

If you look at the Migrations folder in the Database project, you should see the following files:

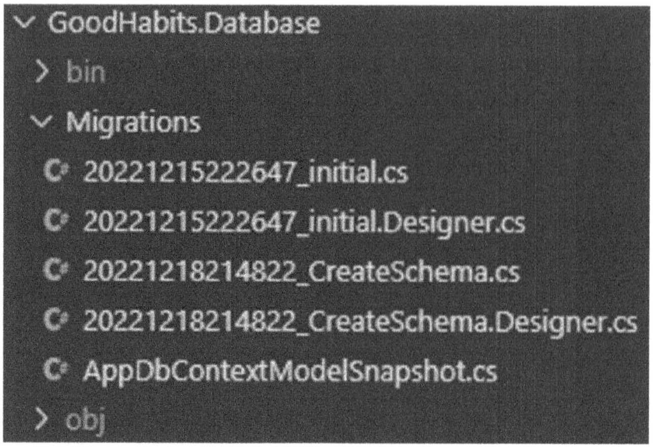

Figure 4.7 – The Migrations folder

In the preceding figure, we can see the `.._InitialSetup.cs` file that contains our first pass, the `.._MultiTenant.cs` file that has the modifications that we made in *Chapter 3*, and the `AdditionalEntities.cs` file that we added in this chapter.

If you recall, in *Chapter 2*, we added the initial migration with the following:

```
dotnet-ef migrations add Initial
```

However, when creating the second migration, we used this instead:

```
dotnet-ef migrations add MultiTenant --startup-project ../GoodHabits.
Api/GoodHabits.Api.csproj
```

The reason for this is that the work we did in *Chapter 3* to introduce the multi-tenancy has added parameters to the constructor, which are defined in the API project. Pointing the migration tool at the API project allows it to create migrations via the `HabitService` project.

If you look into either of these generated classes, you will see two methods, named `Up` and `Down`. These methods allow the migration to be added to the database, or rolled back.

Applying the migrations

Because of how we have configured the databases for multi-tenancy in *Chapter 3*, we do not need to manually update each database. However, should you find you need to manually update a database, you can use the following:

```
dotnet-ef database update --startup-project ../GoodHabits.Api/
GoodHabits.HabitService.csproj
```

Issuing this command will instruct Entity Framework to look in the `Migrations` folder and compare the migrations that it finds there with the migrations that are present in the database. If there are any additional migrations that have not yet been applied to the database, Entity Framework will apply those migrations. Let's take a closer look at the first one we created, called `initial`:

```csharp
using Microsoft.EntityFrameworkCore.Migrations;

#nullable disable

#pragma warning disable CA1814 // Prefer jagged arrays over
                              // multidimensional

namespace GoodHabits.Database.Migrations
{
    /// <inheritdoc />
    public partial class InitialSetup : Migration
    {
        /// <inheritdoc />
        protected override void Up(MigrationBuilder
          migrationBuilder)
        {
            migrationBuilder.CreateTable(
                name: "Habits",
                columns: table => new
                {
                    Id = table.Column<int>(type: "int",
                      nullable: false)
                        .Annotation("SqlServer:Identity",
                          "1, 1"),
                    Name = table.Column<string>(type:
                      "nvarchar(max)", nullable: false),
                    Description =
                      table.Column<string>(type:
                      "nvarchar(max)", nullable: false)
                },
                constraints: table =>
                {
                    table.PrimaryKey("PK_Habits",
                      x => x.Id);
                });

            migrationBuilder.InsertData(
                table: "Habits",
                columns: new[] { "Id", "Description",
                  "Name" },
```

```
                        values: new object[,]
                        {
                            { 100, "Become a francophone",
                              "Learn French" },
                            { 101, "Get really fit",
                              "Run a marathon" },
                            { 102, "Finish your book project",
                              "Write every day" }
                        });
            }

            /// <inheritdoc />
            protected override void Down(MigrationBuilder
              migrationBuilder)
            {
                migrationBuilder.DropTable(
                    name: "Habits");
            }
        }
    }
```

This code is auto-generated by the migration tool, but it's perfectly acceptable to manually adjust the code here if there is a reason to. We can see two methods here. One is called Up, and the other Down.

The Up method creates the table in the database, and the Down method drops the table. This is converted into SQL code, which is issued to the database engine when the database update command is issued.

Summary

We have covered a lot in this chapter!

We learned that data is important in a SaaS application. This can be trying not only from a technical point of view but also from the point of view of a user, as well as the organization building the application.

We then moved on to a technical implementation with Entity Framework, demonstrating how to build a database in code using C#, and then automatically generate migrations and update the database.

We also talked about testing strategies and maintaining a database in production.

In the next chapter, we will build out the API layer and start to interact with the data structure we previously created. The application will start to take shape!

Further reading

- What is Code-First?: `https://www.entityframeworktutorial.net/code-first/what-is-code-first.aspx`

- Testing with Entity Framework: `https://www.michalbialecki.com/en/2020/11/28/unit-tests-in-entity-framework-core-5/`

- Data-driven culture: `https://www.smartkarrot.com/resources/blog/data-driven-culture/`

- Database normalization: `https://www.sqlshack.com/what-is-database-normalization-in-sql-server/`

Questions

1. In what ways can a business monetize data that is present in a SaaS application?
2. What are the ethical considerations around data monetization?
3. How are foreign key relationships represented in Entity Framework?
4. What is the difference between a unit test and an integration test?
5. If I want to roll back migration in a live database, how do I ensure that no data is lost?

5

Building Restful APIs

In today's online-centric digital landscape, **application programming interfaces** (**APIs**) have become ubiquitous in the development of **Software-as-a-Service** (**SaaS**) applications. They allow different systems and applications to communicate with each other and share data. Among the different types of APIs, **Representational State Transfer** (**REST**)ful APIs have become the most widely used and accepted standard, and that is what we will focus on in this chapter.

This chapter will introduce you to the basics of building RESTful APIs and the key principles that guide their design. You will learn about the key components of a RESTful API, such as resources, representations, and the main HTTP verbs (GET, POST, PUT, PATCH, and DELETE).

Additionally, you will learn about the various strategies to version RESTful APIs, such as URL versioning, custom header versioning, media type versioning, and deprecation and sunsetting.

The main topics covered in this chapter are as follows:

- What are RESTful APIs?
- Matching API operations to HTTP verbs
- Designing better with REST
- Versioning public APIs
- Testing APIs

By the end of this chapter, you will have a solid understanding of the key principles and strategies to build RESTful APIs, and you will be well-equipped to design, develop, and test them.

Technical requirements

All code from this chapter can be found at `https://github.com/PacktPublishing/Building-Modern-SaaS-Applications-with-C-and-.NET/tree/main/Chapter-5`.

What are RESTful APIs?

REST was introduced by Roy Fielding in his doctoral dissertation at the University of California, Irvine, in 2000. In his dissertation, Fielding defined the architectural constraints that formed the basis of RESTful systems and described how REST could be used to build scalable and flexible web services. The concepts outlined in his dissertation have since become widely adopted and are used as the foundation to build many modern web APIs.

RESTful APIs are a type of web-based interface that allows for communication between different software systems. They utilize a standard set of constraints and principles defined by the REST architecture to exchange data between a client and server. Resources are identified by unique URLs, and the behavior toward these resources is defined by the HTTP methods used. RESTful APIs are commonly used to build scalable and flexible web services and can return data in different formats, such as JSON or XML. They offer a simple and flexible way for different software systems to interact and exchange data over the internet.

Let's break down what the REST acronym means!

Representational refers to the idea that each resource in a RESTful API is represented by a unique identifier (such as a URL) and can be represented in a variety of formats, such as JSON or XML. The representation of a resource is a snapshot of its current state and can be used by a client to manipulate the resource.

You can think of a resource as an object, such as a description of a user in a system. The user will typically have a unique ID that is used to refer to that user. In a REST system, the user *resource* with an ID = 123 could be *represented* by the following URL:

```
https://www.a-system.com/api/v1/users/123
```

The user can be retrieved, modified, or deleted by using this URL. The URL *represents* the user on any external system that is consuming the **PAI**.

The **state** of a resource refers to its current data and metadata, such as its content, creation date, and any associated metadata. In RESTful APIs, the state of a resource can be manipulated through the use of HTTP methods such as GET, POST, PUT, and DELETE.

If you were to issue a GET request to the preceding dummy URL, you would receive the *state* of the object represented by that URL.

Transfer refers to the transfer of the representation of a resource from a server to a client and vice versa. The transfer is typically performed over the HTTP protocol and is based on the principles of statelessness and uniform resource identification. In RESTful APIs, the transfer of state is used to create, read, update, and delete resources on the server.

RESTful APIs do not have to communicate over HTTP, although they very often do. They could use any other possible communication protocol, such as **remote procedure calls** (**RPCs**). However, the large majority of RESTful APIs use HTTP as the chosen communication mechanism, and that is all that we will consider in this chapter. If you have a use case for an alternative communication protocol, then I hope the information in this chapter is useful in a more general sense!

Before we get into the details of building restful APIs, there are a few general points to consider that will aid our understanding of some of the more complex concepts that will follow.

Idempotency

In the context of a RESTful API, idempotency is a property of an API endpoint that allows multiple identical requests to have the same effect as a single request. This means that, regardless of the number of times the same request is made, the end result should be the same.

An idempotent request will always produce the same response from a server, regardless of how many times it is executed. This property is useful to reduce the chance of errors and conflicts when making multiple requests to the same endpoint, especially when dealing with network connectivity issues or other types of failures.

The most common HTTP methods considered idempotent are GET, PUT, DELETE, and certain types of POST requests. On the other hand, non-idempotent methods, such as POST without specifying the idempotent semantics, may have unintended side effects if repeated multiple times.

This is to say that you can retrieve a resource at a URL as many times as you like, and the response will be the same every time. A GET request is idempotent.

Safety

In the context of a RESTful API, a "safe" operation is one that is guaranteed not to modify the state of a server or have any side effects. Safe operations are read-only and do not alter any resources on the server.

The most common example of a safe operation is a GET request, which retrieves data from the server without changing it. Other safe operations might include OPTIONS, HEAD, and some types of POST requests that are specifically designed to only retrieve data and not make any changes to the server state.

Safe operations are contrasted with "unsafe" operations, such as PUT, POST, and DELETE, operations that modify the state of the server and may have side effects. These operations are considered unsafe because they can have unintended consequences if executed improperly, such as deleting important data or altering resources in unexpected ways.

HTTP status codes

HTTP status codes are three-digit numbers returned by a server in response to a client's request. These codes indicate the outcome of the request and provide information about the status of the requested resource.

There are many HTTP status codes – I will briefly reference only the set that I think is most applicable to building RESTful APIs. Don't worry about memorizing all of these! As you will see, when we start to build up the examples, it is quite intuitive which ones should be used and when! And remember, information like this is only ever a Google search away!

Each status code is a three-digit number. The first digit gives the category of the status code. There are five category codes, each with a specific meaning and purpose. These are as follows:

- **1xx (informational)**: This category of status codes indicates that the request was received and the server is continuing to process it. The most common status code in this category is `100 Continue`.

- **2xx (successful)**: This category of status codes indicates that the request was successfully received, understood, and accepted. The most common status codes in this category are `200 OK` and `201 Created`.

- **3xx (redirection)**: This category of status codes indicates that further action needs to be taken by the client in order to complete the request. The most common status codes in this category are `301 Moved Permanently` and `302 Found`.

- **4xx (client error)**: This category of status codes indicates that the request contains bad syntax or cannot be fulfilled by the server. The most common status codes in this category are `400 Bad Request` and `401 Unauthorized`.

- **5xx (server error)**: This category of status codes indicates that the server failed to fulfill a valid request. The most common status codes in this category are `500 Internal Server Error` and `503 Service Unavailable`.

There are many status codes within these categories. Some of the common and most applicable to RESTful APIs are given here:

- `200 OK`: The request was successful, and the requested resource was returned. This status is commonly returned from a successful GET, PUT, or PATCH request.

- `201 Created`: The request was successful, and a new resource was created as a result. This status code is often returned as the result of a successful POST request.

- `204 No Content`: The request was successful, but no resource was returned. This status is commonly returned from a successful DELETE request.

- `400 Bad Request`: The request was malformed or invalid. With a RESTful API, requests are often in JSON format. This would imply that the object is not correct per the expectations of the API.

- `401 Unauthorized`: The request requires authentication, and the client did not provide valid credentials.

- `403 Forbidden`: The client does not have access to the requested resource.

- `404 Not Found`: The requested resource could not be found. The request is looking for a resource that is not there.

- `405 Method Not Allowed`: The request method, such as `GET`, `POST`, and `DELETE`, is not allowed for the requested resource on the server.

- `500 Internal Server Error`: An unspecified error occurred on the server. This status code is a "catch-all" error to let the user know that something has gone wrong on the server – for example, there could be an exception in the backend code.

- `503 Service Unavailable`: The server is currently unable to handle the request due to maintenance or high traffic.

There are many other HTTP status codes that can be used, each with a specific meaning and purpose. We will make use of these codes when we build out an example later in this chapter.

Dealing with errors

When it comes to dealing with an error code on a RESTful API, it is important to have a clear and consistent approach in place. The HTTP status code is an essential part of this approach, and by using the status codes correctly, clients of the API are able to understand what has gone wrong and will have an idea as to why.

In addition to using the status codes correctly, it also helps to provide clear and informative error messages. These messages should explain what has gone wrong in an easily understood way and, if possible, provide guidance on how to resolve the issue.

The preceding will help users of the API, but it is also very important that the developers of the API are also informed when there has been an error so that they can take action to resolve or prevent recurrences. Because developers cannot watch over every API interaction, this is typically done with logging.

Logging refers to the process of capturing and recording information about an API's behavior and performance and persisting this information in a data store so that it can be searched later to identify issues and troubleshoot problems. Logging is an essential part of any API's operational infrastructure, as it provides a record of what has happened on the system.

This chapter will focus on API implementation, but we haven't forgotten about logging and monitoring – we will cover both in detail in *Chapter 9*!

JSON data formatting

While RESTful APIs do not have to use **JavaScript Object Notation** (**JSON**) formatting, it is an exceptionally popular choice and will be the formatting standard that is used throughout this book. JSON is a lightweight data exchange format that is easy for humans to read and write, and it is also easy for machines to parse and generate. It is completely language-independent, but it does make use of conventions that are familiar to the C family of languages (C, C++, C#, Java, and so on).

Here is an example of some information represented as JSON:

```
{
    "name": "Roger Waters",
    "age": 79,
    "isBassist": true,
    "numbers": [90, 80, 85, 95],
    "address": {
        "street": "123 Main St",
        "city": "A Town",
    }
}
```

JSON data takes the form of key-value pairs, where each key is a string, and each value can be of type string, number, Boolean, null, array, or another JSON object. The ability to nest JSON objects allows for complex types to be represented in this straightforward way.

The keys are always strings, so they are encased in double quotation marks. The values are encased in quotes if they are strings, square brackets if they are arrays, and curly brackets if they are objects. All of these are shown in the preceding snippet.

We have established that JSON-encoded data will be sent and received over HTTP. Next, we will look at how that data is transmitted, looking at the most common HTTP verbs and describing how and when to use them.

Matching API operations to HTTP verbs

In language, a verb is a "doing" word. It describes an action, state, or occurrence. In English, examples of verbs include "run," "think," "sing," and "do," as well as many thousands more!

The HTTP verbs describe things that you can "do" over HTTP! Five main verbs are used – GET, POST, PUT, PATCH, and DELETE. Each of these serves a different purpose, although the precise purpose of each is not tightly defined, and it is not uncommon to accidentally use the wrong one. In this section, we will cover the uses of the five commonly used HTTP verbs, and we will give an example of them being used in our demo application.

GET

The GET HTTP verb is used to *retrieve a resource from a server*. When a client sends a GET request to a server, the server responds by returning the requested resource to the client. The resource can be any type of data, such as a web page, image, or file. The GET verb is the most widely used HTTP verb and is considered to be a safe and idempotent method, which means that it can be called multiple times without any side effects. It is also cacheable, which means that the response can be stored in a cache and reused to improve performance. The GET verb should only be used to retrieve information and never to make changes on the server.

When it comes to RESTful APIs, the GET verb is used to retrieve a representation of a resource or a collection of resources from a server. The resource can be identified by a **Uniform Resource Identifier (URI)**, and the representation of the resource is typically in a format such as JSON or XML. The GET request can also include query parameters, which can be used to filter the results or specify the format of the returned data. The server responds to a GET request with a representation of the resource in the form of an HTTP response, along with the appropriate HTTP status code. The most common status code for a GET request is 200 OK, indicating that the request was successful and the resource was returned.

POST

The POST HTTP verb is used to *submit an entity to be processed by the resource identified by the URI*. A POST request is typically used to create a new resource or (sometimes) to update an existing one. The POST verb is not idempotent, which means that it can have different effects depending on how many times it is called. It is also not safe, meaning that it can modify the resource on the server.

When a client sends a POST request to a RESTful API, typically, the server creates a new resource with the data provided in the request body and returns a response with a status code indicating the outcome of the request. The most common status code for a successful POST request is 201 Created, indicating that a new resource has been successfully created. The URI of the newly created resource is typically included in the response headers so that the client application can retrieve and work with the newly created resource immediately. The data in a POST request can be in any format, such as JSON or XML, but it is usually in the JSON format.

DELETE

The DELETE HTTP verb is used to *delete a resource from a server*. A DELETE request is used to remove the specified resource from the server. The DELETE verb is idempotent, which means that it can be called multiple times without any side effects. It is also not safe, meaning that it can modify the resource on the server.

When a client sends a DELETE request to a RESTful API, the server deletes the specified resource and returns a response with a status code, indicating the outcome of the request. The most common status code for a successful DELETE request is 204 No Content, indicating that the resource has been successfully deleted. The client usually doesn't receive any content in the response body, only the status code. The DELETE request usually requires the URI of the resource to be specified in the request so that the server is able to identify which resource to delete.

It's worth noting that some RESTful APIs may not allow DELETE requests and will return a 405 Method Not Allowed status code if a DELETE request is received.

PUT

The PUT HTTP verb is used to *update an existing resource or create a new one if it does not exist*. A PUT request is used to submit a representation of the resource to be updated or created. The representation of the resource is included in the request body and typically encoded in a format such as JSON or XML. The PUT verb is idempotent, which means that it can be called multiple times without any side effects. It is also not safe, meaning that it can modify the resource on the server.

When a client sends a PUT request to a RESTful API, the server updates the specified resource with the data provided in the request body and returns a response with a status code, indicating the outcome of the request. The most common status code for a successful PUT request is 200 OK, indicating that the resource has been successfully updated. If a new resource is created, the 201 Created status code will be returned. The URI of the updated resource is typically included in the response headers.

It's worth noting that PUT requests may require a client to send the full representation of the resource in the request body, including all properties, even if only a few of them are to be updated. This can make a PUT request inefficient in terms of bandwidth used, and it may be better to use the PATCH verb.

PATCH

The PATCH HTTP verb is used to *partially update an existing resource on a server*. A PATCH request is used to submit a set of changes to be made to the specified resource rather than replacing the entire resource. The set of changes is typically encoded in a format such as JSON or XML and is included in the request body. The PATCH verb is idempotent, which means that it can be called multiple times without any side effects. It is also not safe, meaning that it can modify the resource on the server.

In a RESTful API context, the PATCH verb is typically used to partially update an existing resource on the server. When a client sends a PATCH request to a RESTful API, the server applies the changes provided in the request body to the specified resource and returns a response with a status code indicating the outcome of the request. The most common status code for a successful PATCH request is 200 OK, indicating that the resource has been successfully updated. The URI of the updated resource is typically included in the response headers. The data in a PATCH request can be in any format, such as JSON or XML.

It's worth noting that PATCH requests require a client to send a specific set of changes to be made to the resource rather than the full representation of the resource. This makes PATCH requests more lightweight and efficient than PUT requests for partial updates.

Data transfer objects

Another important concept to understand when working with APIs is **data transfer objects (DTOs)**. A DTO is a design pattern that is commonly used to transfer data between layers or systems. In the case of a RESTful API, this is typically to transfer data from the backend (API) to the frontend **user interface** (**UI**). The general purpose of DTOs is to decouple the structure of the data from the underlying systems that use it, allowing for more flexibility and easier maintenance. They also provide a standardized way to handle data, making it easier for different components of a system to communicate and exchange information.

DTOs are particularly useful in RESTful APIs, as they provide a standard way to represent data when sending and receiving requests between a client and a server. When designing a RESTful API, the use of DTOs allows the API to define the structure of the data that is exchanged without having to tightly couple the implementation of the API to the structure of the data. This decoupling makes it easier to evolve the API and make changes to the underlying data model without affecting the API clients. Additionally, using DTOs enables the API to shape the data it returns to better match the needs of the client, reducing the amount of data transferred over the network and improving performance. Furthermore, DTOs can be used to validate the data being passed between the client and server, ensuring that only valid data is accepted and processed.

The first set of DTOs that we will see in this chapter will look a lot like the entity types that we defined for our database in *Chapter 3* and *Chapter 4*, and they will relate to operations that we may wish to perform on the database. For example, the following entity type represents Habit in the database:

```
[Index(nameof(Id), nameof(UserId))]
public class GoodHabit : IHasTenant
{
    public int Id { get; set; }
    public string Name { get; set; } = default!;
    public int UserId { get; set; }
    public virtual User User { get; set; } = default!;
    public virtual ICollection<Progress> ProgressUpdates {
        get; set; } = default!;
    public virtual ICollection<Reminder> Reminders { get;
        set; } = default!;
    public virtual Goal Goal { get; set; } = default!;
    public Duration Duration { get; set; }
    public string TenantName { get; set; } = default!;
}
```

Let's say that we wanted to create a simple `Habit` that only had a `Name` property populated and was tied to a certain `User`. We could send the following DTO:

```
public class CreateHabitDto {
    public string Name { get; set; }
    public int UserId { get; set; }
}
```

This could be used by the backend to create a simple `GoodHabit` object in the database.

If we wanted to retrieve the `GoodHabit` objects but only with the name and ID properties, we could use a DTO that looked like this:

```
public class GetGoodHabitDto {
    public int Id { get; set; }
    public string Name { get; set; }
}
```

And, if we needed more information than simply a name and an ID, we could further define another DTO that looked like this:

```
public class GetGoodHabitDetailDto {
    public int Id { get; set; }
    public string Name { get; set; }
    public string UserName { get; set; }
    public string GoalName { get; set; }
    public string Duration { get; set; }
}
```

You can see how we can start with the entity type that has a lot of database-specific information, and we can selectively model that data in different ways for different use cases.

We will illustrate this point with examples later in this chapter!

Designing better with REST

Good design with RESTful APIs is essential to create an API that is easy to use and understand. One of the key principles of REST is that it is based on the use of resources and their representations. Therefore, it's essential to design the API's resources and their representations in a way that is meaningful and consistent.

When designing resources, it's important to use URIs that are meaningful and consistent. Resources should be named in a way that is easy to understand, and the URIs should be structured logically and hierarchically – for example, `/users/1/orders/2` is more meaningful than `/users?id=1&orderid=2`.

Representations of resources should be in a format that is easy to parse, such as JSON or XML. It's also important to use the `Content-Type` and `Accept` headers to specify the format of the request and response. This allows the client to specify its preferred format and the server to provide the appropriate representation.

Another important aspect of a good RESTful API design is stateless communication. This means that a client and server should not maintain any state between requests. This can be achieved through the use of HTTP headers and cookies. This allows for a higher degree of scalability and flexibility, as the server does not have to maintain any state for each client.

Security is also an important aspect of RESTful API design. It's important to use secure communication protocols, such as HTTPS, and to implement authentication and authorization mechanisms to protect an API's resources.

In addition to the aforementioned points, good RESTful API design also includes error handling, caching, versioning, and documentation. Good documentation is essential for developers to understand how to use the API. It's also important to provide clear and informative error messages to clients when an error occurs. In the previous chapter, we built up a database. In this chapter, we'll now add an API to interact with the database that we built. We will use a very typical layer structure for a SaaS app that looks something like this:

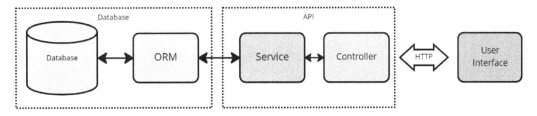

Figure 5.1 – The layering structure

In the preceding diagram, we can see the database and the **object–relational mapping** (**ORM**) that we covered in previous chapters. In this chapter, we are learning about the API, so we will build a service layer to interact with the database and a controller layer to handle communication with the UI or any other client.

A service layer is an architectural pattern in software design that acts as an intermediary between the application's UI and the underlying data storage. The main purpose of a service layer is to encapsulate and abstract the business logic of an application, promoting the separation of concerns and making it easier to maintain and modify the code. It also enables better unit testing, as the service layer can be tested in isolation from the rest of the application. Additionally, it can improve the scalability of an application by allowing the UI and data storage components to evolve independently.

In a RESTful API, controllers are responsible for handling HTTP requests from clients and returning appropriate HTTP responses. They act as an intermediary between a client and an application's business logic, using the appropriate service layer to perform any necessary operations.

Controllers are responsible for mapping URLs to specific actions in an application, such as retrieving data from a database, creating new resources, or updating existing ones. They parse an incoming request to determine the desired action and then use the appropriate service layer to perform that action and generate a response.

An example API design

You will recall that in *Chapter 3*, we started to build out the `HabitService` API with a couple of endpoints. We'll start from where we left off in *Chapter 3*, but we will add a lot more functionality to the controller!

The three endpoints that we have already added are the following:

A `GET` endpoint that gets a single habit based on a passed-in ID:

```
public async Task<IActionResult> GetAsync(int id) => Ok(await _
habitService.GetById(id));
```

Another `GET` endpoint that returns all of the habits:

```
public async Task<IActionResult> GetAsync() => Ok(await _habitService.
GetAll());
```

And finally, a `POST` endpoint that will create a new habit in the database:

```
public async Task<IActionResult> CreateAsync(CreateHabitDto request)
=> Ok(await _habitService.Create(request.Name, request.Description));
```

In this section, we will add an endpoint for each of the five primary HTTP verbs that we have discussed in this chapter. We already have `GET` and `POST`, so we will add `PUT`, `PATCH`, and `DELETE`.

DTOs

But, before we write the endpoints, we will first add the DTOs. We already have a `CreateHabitDto` that was added in *Chapter 3*. Run the following script from the root folder, or add the files manually:

```
cd GoodHabits.HabitService/Dtos; \
touch HabitDetailDto.cs; \
touch HabitDto.cs; \
touch UpdateHabitDto.cs; \
cd ..;
```

Copy the following into the `HabitDetailDto` class:

```
namespace GoodHabits.HabitService.Dtos;
public class HabitDetailDto {
    public int Id { get; set; }
    public string Name { get; set; } = default!;
```

```
    public string UserName { get; set; } = default!;
    public string GoalName { get; set; } = default!;
    public string Duration { get; set; } = default!;
}
```

Then add the following to the `HabitDto` class:

```
namespace GoodHabits.HabitService.Dtos;
public class HabitDto
{
    public int Id { get; set; } = default!;
    public string Name { get; set; } = default!;
    public string Description { get; set; } = default!;
}
```

And finally, add the following to the `UpdateHabitDto` class:

```
namespace GoodHabits.HabitService.Dtos;
public class UpdateHabitDto
{
    public string Name { get; set; } = default!;
    public string Description { get; set; } = default!;
}
```

That is all that is required for the DTOs. We will use these when we start to build out the endpoints.

AutoMapper

We now need to consider a tool that we will use to convert between the database types and the DTOs. That tool is AutoMapper.

AutoMapper is an open source library that enables you to establish a configuration to convert one object to another. This can be particularly useful when translating between different types of objects, such as database entities and DTOs. Even though both the DTO and the entities might have a similar structure, their implementation can differ. This library helps to keep your codebase clean and maintainable by reducing the amount of repetitive and boilerplate mapping code that is needed to translate between different types, making your application more efficient and easier to modify or add new features.

AutoMapper is designed to make it easy to map one type of object to another, and it provides a simple, fluent API to define the mappings. Some of the key features of AutoMapper include:

- Support for flattening and unflattening of object hierarchies
- Support for converting between different data types and custom-type converters
- Support for advanced configuration options, such as mapping to and from interfaces and inheritance hierarchies

- Support for custom logic and conventions to be used during mapping

- Support for **Language Integrated Query (LINQ)** expressions to define mappings

AutoMapper can help to keep your codebase clean and maintainable by reducing the amount of repetitive, boilerplate mapping code that is required to convert between different types. This can make your application more efficient, and it makes it easier to add new features or make changes to existing ones.

To get started with AutoMapper, install the tool in the API project with the following command:

```
dotnet add package AutoMapper.Extensions.Microsoft.DependencyInjection
```

With the packages updated, as shown in the preceding code snippet, we can start to create mappings for the database types that we have created.

You will recall that we added an entity type called 'Habit' in the database project in *Chapter 3*, and we have added a number of additional properties to the object in *Chapter 4*. If you run the `HabitService` and use Thunder Client to hit the endpoint that returns all of the habits, you will see that the data which is returned includes all of these additional properties.

This habit class represents a database entity. It is very specific to the database and works very well to represent the idea of a good habit for that specific use case. But, it does not work well to transmit data to the UI.

We would rather that the data was sent to the UI in the form of a DTO, such as the one we created previously.

- There is no need to include the collections for the progress updates or the reminders. Including this information could add a huge amount to the required bandwidth of the app.

- The `TenentName` property is of no use to the user because they will already know which tenant they are!

The DTO that we have created looks like this:

```
namespace GoodHabits.HabitService.Dtos;
public class HabitDto
{
    public int Id { get; set; } = default!;
    public string Name { get; set; } = default!;
    public string Description { get; set; } = default!;
}
```

In this example, we are taking the ID, name, and description of the habit directly from the entity type, but more complex transformations are also possible.

While we could simply copy the properties over manually, that could quickly get very tedious, so we'll use `AutoMapper` to do this automatically!

Start by going into the `Program.cs` class and adding the `AutoMapper` service:

```
builder.Services.AddAutoMapper(AppDomain.CurrentDomain.
GetAssemblies());
```

Next, open up the `HabitsController` class, and add the following to the `using` statements:
using `AutoMapper`.

Then, add the following to the class definition:

```
private readonly IMapper _mapper;
```

Next, add the `using` statement:

```
using AutoMapper;
```

Next, modify the constructor to take in the mapper, like so:

```
public HabitsController(
    ILogger<HabitsController> logger,
    IHabitService goodHabitsService,
    IMapper mapper
    )
{
    _logger = logger;
    _habitService = goodHabitsService;
    _mapper = mapper;
}
```

Finally, modify the two existing GET endpoints in the controller to use `AutoMapper`, like this:

```
[HttpGet("{id}")]
public async Task<IActionResult> GetAsync(int id) =>
  Ok(_mapper.Map<HabitDto>(await
  _habitService.GetById(id)));

[HttpGet]
public async Task<IActionResult> GetAsync() =>
  Ok(_mapper.Map<ICollection<HabitDto>>(await
  _habitService.GetAll()));
```

Before, the controller was simply returning the entity object from the database, now that object is being mapped automatically to a DTO, which is returned from the controller. This does require a little configuration, though.

The final step is to tell `AutoMapper` how it should convert between the two types.

Add a folder called `Mappers` in the `HabitService` project and a class called `HabitMapper.cs`. You can use this script:

```
mkdir Mappers; \
cd Mappers; \
touch HabitMapper.cs; \
cd ..;
```

In this class, add the following:

```
using AutoMapper;
using GoodHabits.HabitService.Dtos;
using GoodHabits.Database.Entities;

namespace GoodHabits.HabitService.Mappers;

public class HabitMapper : Profile
{
    public HabitMapper()
    {
        CreateMap<Habit, HabitDto>();
    }
}
```

The `CreateMap` method instructs `AutoMapper` to map between the two types.

You can now hit the endpoint using Thunder Client to get the habits, and you should see something like this:

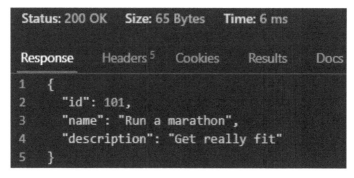

Figure 5.2 – A successful response

This section has demonstrated how we can automatically convert between database types and types for data transfer. This is a very important piece of the API puzzle, and understanding how to operate AutoMapper will help you write better code and also reduce the amount of data sent between the API and any connected clients.

Modify the service

Before we can build the additional endpoints on the API to update and delete the habits, we need to add some functionality to the service class. We already created the service class and interface in *Chapter 3*, but we will extend the functionality here.

Add the following to the interface:

```
using GoodHabits.Database.Entities;
using GoodHabits.HabitService.Dtos;

namespace GoodHabits.HabitService;
public interface IHabitService
{
Task<Habit> Create(string name, string description);
Task<Habit> GetById(int id);
Task<IReadOnlyList<Habit>> GetAll();
Task DeleteById(int id);
Task<Habit?> UpdateById(int id, UpdateHabitDto request);
}
```

The `HabitService` class that implements the preceding interface will need to have two methods added to delete and update the habits that are stored in the database. Add the following two methods to the `HabitService` class:

```
    public async Task DeleteById(int id)
    {
        var habit = await _dbContext.Habits!.FindAsync(id)
          ?? throw new ArgumentException("User not found");
        _dbContext.Habits.Remove(habit);
        await _dbContext.SaveChangesAsync();
    }

    public async Task<Habit?> UpdateById(int id,
      UpdateHabitDto request)
    {
        var habit = await _dbContext.Habits!.FindAsync(id);
        if (habit == null) return null;

        habit.Name = request.Name;
        habit.Description = request.Description;
        await _dbContext.SaveChangesAsync();
        return habit;
    }
```

You will also need to add a `using` statement to the service class:

```
using GoodHabits.HabitService.Dtos;
```

That is all that is required in the service layer.

That has been quite a lot of configuration, but we are now ready to build the controller class.

Add to the controller

We have done most of the heavy lifting already by adding the DTOs, configuring AutoMapper, and building the service layer. We will need to add three additional endpoints to the controller. Let's start with the DELETE endpoint:

```
[HttpDelete("{id}")]
public async Task<IActionResult> DeleteAsync(int id)
{
    await _habitService.DeleteById(id);
    return NoContent();
}
```

This is pretty straightforward. It uses the service method to delete the entry in the database and then returns `NoContent` – which is considered best practice for a delete method.

Next, add the endpoint to update the object with the PUT verb:

```
[HttpPut("{id}")]
public async Task<IActionResult> UpdateAsync(int id,
  UpdateHabitDto request)
{
    var habit = await _habitService.UpdateById(id,
      request);
    if (habit == null)
    {
        return NotFound();
    }

    return Ok(habit);
}
```

There is some error trapping here, which returns 404 if the client attempts to update an entry that does not exist.

Finally, add the endpoint that updates an object using the PATCH verb:

```
[HttpPatch("{id}")]
public async Task<IActionResult> UpdateAsync(int id,
```

```
      [FromBody] JsonPatchDocument<UpdateHabitDto> patch)
  {
      var habit = await _goodHabitsService.GetById(id);
      if (habit == null) return NotFound();

      var updateHabitDto = new UpdateHabitDto { Name =
        habit.Name, Description = habit.Description };
      try
      {
          patch.ApplyTo(updateHabitDto, ModelState);
          if (!TryValidateModel(updateHabitDto)) return
            ValidationProblem(ModelState);
          await _goodHabitsService.UpdateById(id,
            updateHabitDto);
          return NoContent();
      }
      catch (JsonPatchException ex)
      {
          return BadRequest(new { error = ex.Message });
      }
  }
```

This is a little more involved, as it uses `JsonPatchDocument` to modify the object. You will also need to add two `using` statements:

```
using Microsoft.AspNetCore.JsonPatch;
using Microsoft.AspNetCore.JsonPatch.Exceptions;
```

That is all that we need to do at this stage. We now have a good example of the five most common HTTP verbs. Before we move on, we should test that these all work. We will use Thunder Client for this.

Testing

To test the endpoint that we have just added, we will need a test client. In keeping with the theme of using **Visual Studio Code** (**VS Code**), we will add an extension to the code so that we can do everything in one place. We have touched on this tool a couple of times already, but we will take a close look in this section.

You'll see the Thunder Client icon on the extensions toolbar:

Figure 5.3 – The Thunder Client icon

With Thunder Client, you can hit your API straight from VS Code and check that it is behaving as expected. We'll do this now. Start the API running by going to the terminal in VS Code, navigating to the API project, and typing the following:

```
dotnet run
```

This will build the project and get the API running. Now we can start adding the tests!

Add a GET request

Now, complete the following steps:

1. Click on the Thunder Client icon (*if you don't see this on the left-hand menu, exit and restart the Docker environment*).

2. Click on the **Collections** tab and add a new collection called `GoodHabits`.

3. *Right-click* on the newly created collection, click **New Request** and add a request called `GET` Habits (don't click on the bit below the **New Request** button; right-click the collection).

    Your collection should look like this:

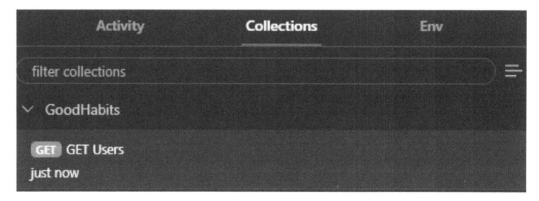

Figure 5.4 – The Thunder Client collection

4. Click on the **GET** user request:

    A. Set the URL to `http://localhost:5100/api/habits`.

    B. In the **Headers** tab, add a `tenant` key with the `CloudSphere` value (you'll remember from *Chapter 3* that we need to specify the tenant for the multi-tenancy).

When you're done, it should look like this:

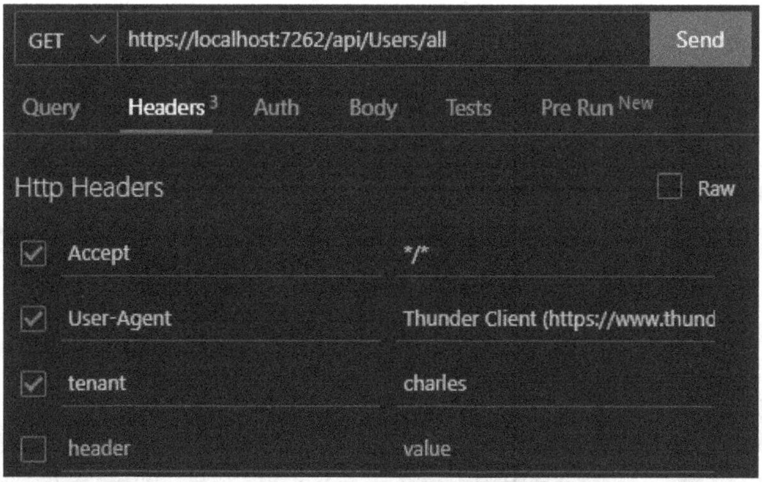

Figure 5.5 – The configured request

The preceding screenshot shows a correctly configured GET request that should return all of the habits in the database.

Finally, click the **Send** button to issue the GET request and test the endpoint. You will see the following:

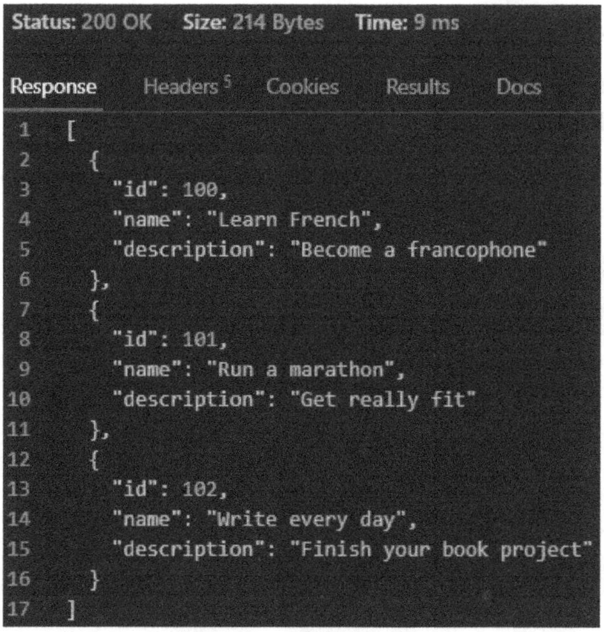

Figure 5.6 – The returned habits

We have put in quite a lot of work to get to this stage! We are showing the data from the `SeedData` file in the database project, returned from our `HabitsService`. We will shortly build a UI to present this information.

Add a POST request

Repeat the preceding, building a `POST` user request. In this case, we'll need to specify the habit details in the body in JSON format:

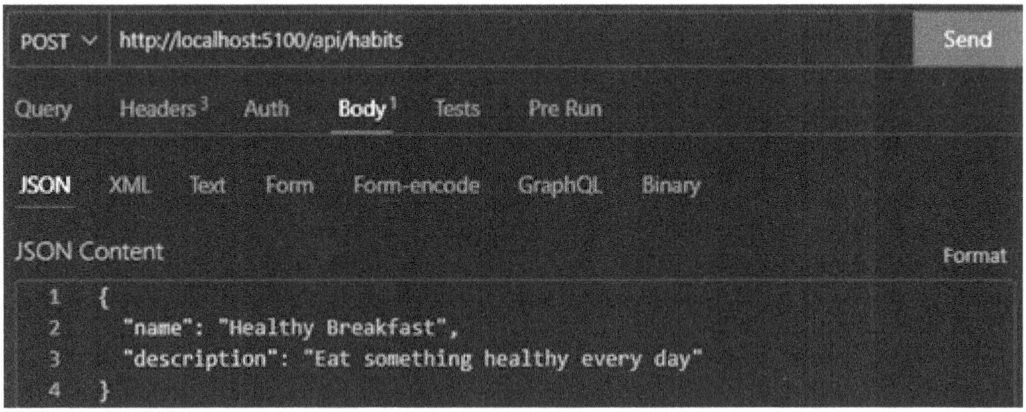

Figure 5.7 – The configured POST request

You can see that the JSON specified matches the `CreateHabitDto` class.

Don't forget to set the tenant in the header and change the request type to `POST`! Hitting **Send** will confirm that the habit has been created.

So far, we have tested the `get all` endpoint and the `POST` endpoint. It would be a useful exercise to add another `GET` to test the `get-by-id` endpoint as well!

Add a DELETE request

We may want to delete a habit from the database, so we have added the required methods to the service and the controller. We can test this again in the same way:

DELETE
http://localhost:5100/api/habits/100
Send
Query
Headers 3
Auth
Body
Tests
Pre Run
JSON
XML
Text
Form
Form-encode
GraphQL
Binary
JSON Content
Format
1

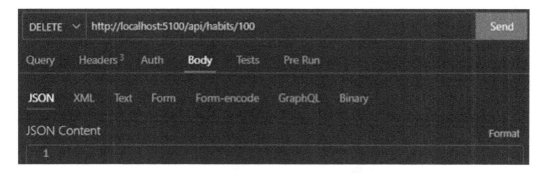

Figure 5.8 – The configured DELETE request

The preceding screenshot shows that no content is required in the body. But don't forget to add the tenant header!

Add a PUT request

Testing the `PUT` endpoint that we have added is fairly straightforward. Configure a `PUT` request like this:

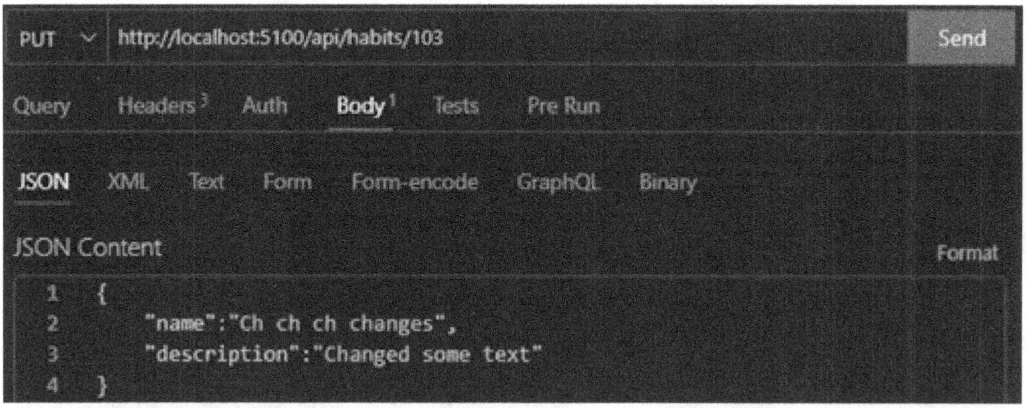

PUT
http://localhost:5100/api/habits/103
Send
Query
Headers 3
Auth
Body 1
Tests
Pre Run
JSON
XML
Text
Form
Form-encode
GraphQL
Binary
JSON Content
Format
1 {
2 "name":"Ch ch ch changes",
3 "description":"Changed some text"
4 }

Figure 5.9 – The configured PUT request

The preceding figure shows how to configure a `PUT` request. This will alter the name and description of the habit with `id=103`. You may need to change the ID in the URL if you have made changes to the data along the way. You can check that this has made the changes by hitting the `get-by-id` endpoint again.

Add a PATCH request

Testing the PATCH endpoint is a little more tricky. You will recall that the PATCH endpoint that we built in the controller is expecting a JsonPatchDocument object, so this is what we will have to supply. A Patch document could look like this:

```
[
    {
        "op": "replace",
        "path": "/Name",
        "value": "A New Name"
    }
]
```

The preceding code uses the replace operator to change the value of the Name variable. We can set up the request like this:

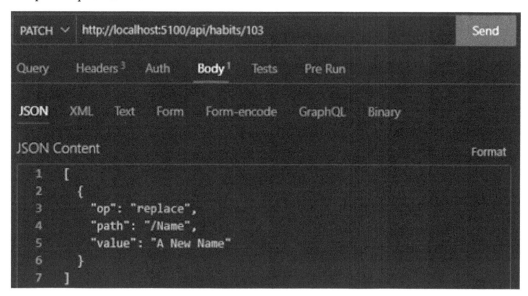

Figure 5.10 – The configured PATCH request

Configuring this and hitting send will update the habit with id=103. Again, you may need to change the ID in the URL.

You will notice that the service will return a 204 No Content response upon a successful patch. This is the expected behavior and is considered best practice for the response to a successful PATCH.

This would be a good time to talk a bit more about HTTP codes!

Using the correct HTTP codes

Earlier in this chapter, we talked about HTTP status codes, and we looked at a few that may be useful to consider, even for the very basic examples we saw previously.

The two that we should add and test are as follows:

- If a request is made to get a user that does not exist, the correct status code should be 404 Not found

- If a successful request is made to create a new user, the correct status code should be 201 Created

Because we have separated the service and the controller, we do not have to change any of the service logic to facilitate this. The controller has the sole responsibility for assigning the HTTP status codes. While the example here is fairly straightforward, I hope you can see how separating the logic in this way can be very beneficial when things start to get more complex.

We'll start by modifying the GET method, which takes an id parameter to look like this:

```
[HttpGet("{id}")]
public async Task<IActionResult> GetAsync(int id)
{
    var habit = await _habitService.GetById(id);
    if (habit == null) return NotFound();
    return Ok(_mapper.Map<HabitDto>(await
      _habitService.GetById(id)));
}
```

We have simply added a check to see whether the user object is null, and if so, we return NotFound(), which will return the 404 status code.

You can test this in Thunder Client by requesting a user ID that you know doesn't exist in your database:

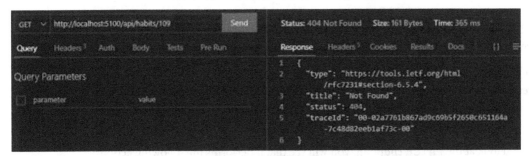

Figure 5.11 – Habit not found

In the preceding screenshot, we have demonstrated that requesting a non-existent user now results in a 404 HTTP status code.

Next, let's fix the HTTP code to create a new user. Modify the Create endpoint to look like the following:

```
[HttpPost]
public async Task<IActionResult>
  CreateAsync(CreateHabitDto request)
{
    var habit = await _habitService
      .Create(request.Name, request.Description);
    var habitDto = _mapper.Map<HabitDto>(habit);
    return CreatedAtAction("Get", "Habits", new { id =
      habitDto.Id }, habitDto);
}
```

We have changed the return from Ok() to CreatedAtAction(...). This will return 201 – Created and also the location of the newly created resource to the user.

If you go back into Thunder Client and create another user, you will see the following:

Figure 5.12 – Created with a 201 status code

Clicking on the **Headers** tab will give you the location of the newly created resource. This can be very useful to consumers of your API who may want to interact with the new resource immediately.

If you look at the PUT and the PATCH endpoints in HabitsController, you can see that we are returning a few different status codes depending on what has happened when processing the response. It would be a good exercise to go through these and convince yourself that you have understood why each of them has been selected.

If you look at the PATCH endpoint, you will see that it performs the following actions:

- It checks to see whether the ID provided is valid and, if not, returns 404 – Not Found

- It checks to see whether the updated model is valid and, if not, returns a validation problem (a subset of 400 Bad Request)

- If there are any other issues with the update, it returns `400 Bad Request`

- If there are no issues, it returns `204 No Content`

The `No Content` HTTP status code (`204`) is used to indicate that the server has successfully processed the request and that there is no response body to return. In the case of a `PATCH` request, the `No Content` status code is used to indicate that the server has successfully processed the update to the resource without returning any content in the response. The idea is that the client already knows what the updated resource looks like, and therefore, there is no need to return the updated resource information in the response. The client can simply assume that the update was successful and that the resource was updated as requested.

Versioning public APIs

Versioning public RESTful APIs is the process of creating and maintaining multiple versions of an API to accommodate changes and new features. This ensures that existing clients are not affected by changes made to the API and that new clients can take advantage of the new features.

Versioning is a critical aspect of API development and maintenance because it enables the evolution of an API over time while maintaining compatibility with existing clients. This is particularly important in cases where APIs are used by multiple clients, and breaking changes would impact the functionality of those clients. With versioning, multiple versions of an API can coexist, and clients can choose to upgrade to the latest version or continue using an earlier version that suits their needs. This way, versioning provides the necessary flexibility for APIs to evolve and improve over time without disrupting the stability of existing integrations.

There are several strategies to version RESTful APIs, each with its own advantages and disadvantages:

- **URL versioning**: This strategy involves including the version number in the URI of the API, such as `/v1/users` or `/v2/users`. This approach is easy to implement and understand, but it can be difficult to maintain and scale as the number of versions increases.

- **Custom header versioning**: This strategy involves including the version number in a custom header, such as `X-API-Version`. This approach allows for more flexibility, as the URI does not have to change, but it can be more complex to implement and may not be supported by all clients.

- **Media type versioning**: This strategy involves using different media types to represent different versions of the API, such as `application/vnd.example.v1+json` or `application/vnd.example.v2+json`. This approach allows for more flexibility, as the URI and headers do not have to change, but it can be more complex to implement and may not be supported by all clients.

- **Deprecation and sunsetting**: This strategy involves marking old versions of the API as deprecated and eventually sunsetting them. This approach allows for a gradual transition and gives clients time to update their code before the old version is removed.

It is worth noting that the most appropriate versioning strategy will depend on the specific needs of an API and its clients. It's important to communicate the versioning strategy and the timeline for the deprecation of old versions to the API's clients in order to minimize disruption and allow them to plan accordingly.

The most common way to version an API is to include the version number in the URL of the API endpoint. For example, the URL of an API endpoint might look like this:

```
https://api.example.com/v1/resources
```

This method allows different versions of an API to coexist and makes it easy to manage changes to the API over time by simply changing the version number in the URL. This also allows clients to choose which version of the API they want to use in their applications, and it helps prevent breaking changes in the API from affecting existing client applications.

If a second version of the preceding example was created, it could be found at the following link:

```
https://api.example.com/v2/resources
```

A huge benefit of this is that both versions can exist at the same time and users who have clients that still expect the v1 version can continue to work seamlessly. Of course, supporting multiple versions can be hard, and ideally, this would be a transient state with the intention to deprecate the v1 version at some point.

Example code showing how to version an API

Previously in this chapter, we built a controller to manage the users and added a number of endpoints to it. We have not yet added any versioning to the API, though; note that the URLs that we have tested with (using Thunder Client) do not have a version associated with them, such as the following:

```
http://localhost:5100/api/habits
```

Let's change that!

Start by opening a console and adding the versioning package to the HabitService project:

```
dotnet add package Microsoft.AspNetCore.Mvc.Versioning
```

Add the using statement into Program.cs:

```
using Microsoft.AspNetCore.Mvc.Versioning;
```

Next, copy the following into Program.cs:

```
builder.Services.AddApiVersioning(opt =>
    {
        opt.DefaultApiVersion = new
```

```
        Microsoft.AspNetCore.Mvc.ApiVersion(1,0);
    opt.AssumeDefaultVersionWhenUnspecified = true;
    opt.ReportApiVersions = true;
    opt.ApiVersionReader = ApiVersionReader.Combine(new
      UrlSegmentApiVersionReader(),
      new HeaderApiVersionReader("x-api-version"),
      new MediaTypeApiVersionReader("x-api-version"));
});
```

Let's review the preceding code in detail:

- The first flag sets the default API version. This allows a client to work with the API without specifying a version.

- The second flag instructs the application to use the default if nothing is specified. This is an example of defensive programming – your users will thank you for this!

- The third flag is returns options – this returns the available versions in the response header so that a calling client can see that there are options available for that method.

- Finally, `ApiVersionReader` makes it possible for clients to choose whether to put the version in the URL or the request header. Again, it's good to give consumers of the API the choice in this.

Now, we need to update `HabitsController` to work with multiple versions.

To illustrate this point, we'll just use a simple GET endpoint. But you can apply the same logic to any of the endpoints.

Change the attributes of the `HabitsController` class to the following:

```
[ApiController]
[Route("api/[controller]")]
[Route("api/v{version:apiVersion}/[controller]")]
[ApiVersion("1.0")]
```

Let's prove what we have done by adding an endpoint to the controller and mapping it to `version 1.0`, like this:

```
[MapToApiVersion("1.0")]
[HttpGet("version")]
public virtual async Task<IActionResult> GetVersion()
{
    return Ok("Response from version 1.0");
}
```

We have flagged this method as virtual so that we can override it in a subsequent version.

Create a file called `HabitsControllerv2.cs` and add the following to it:

```
using Microsoft.AspNetCore.Mvc;

namespace GoodHabits.HabitService.Controllers.v2;

[ApiController]
[Route("api/[controller]")]
[Route("api/v{version:apiVersion}/[controller]")]
[ApiVersion("2.0")]
public class HabitsController : ControllerBase
{
    [MapToApiVersion("2.0")]
    [HttpGet("version")]
    public virtual IActionResult GetVersion()
    {
        return Ok("Response from version 2.0");
    }
}
```

Note that this maps the `version` endpoint to the v2 API. You can test this in Thunder Client in the usual way, and you will see that changing the version you provide in the URL changes which response you get.

Also, note that we have specified the `Route` attribute twice – once with the version included and once without. This allows the default versioning that we specified in `Program.cs` to take effect.

In Thunder Client, run three tests – one test without a version, one with v1, and one with v2:

- No version: `http://localhost:5100/api/Habits/version`
- Version 1: `http://localhost:5100/api/v1/Habits/version`
- Version 2: `http://localhost:5100/api/v2/Habits/version`

You will see that the first one returns v1, as that is the default, and you will see that the other two perform as you would expect.

You should also note that the requests that we previously set up in Thunder Client continue to operate as expected. This is great from the point of view of a consumer of the API. We have just introduced versioning and added v2 without breaking any existing functionality!

Testing APIs

In this chapter, we have demonstrated quite extensively how to test your API using Thunder Client. Testing APIs (and testing in general) is a huge subject that could be the subject of a book on its own. If you are interested, I have provided some pointers for further reading in the following section!

The following list provides some examples of the type of testing that you may want to carry out to ensure that your API is functioning correctly. Unit testing involves testing individual components of an API to ensure that they are working as expected. This is typically done using a unit testing framework, such as NUnit, and can be automated:

- Functional testing involves testing an API end to end to ensure that all the components are working together correctly. This can be done manually or by using an automated testing tool, such as Selenium or TestComplete.

- Integration testing involves testing an API in conjunction with other systems, such as a database or other API. This can be done using an integration testing framework, such as Cucumber or FitNesse.

- Performance testing involves testing an API to ensure that it can handle the expected load and is performing optimally.

- Security testing involves testing an API to ensure that it is secure and not vulnerable to common security threats, such as SQL injection or cross-site scripting. This can be done using a security testing tool, such as Nessus or OWASP ZAP.

- Usability testing involves testing an API to ensure that it is easy to use and understand. This can be done manually or by using a usability testing tool, such as UserTesting or Crazy Egg.

- Postman is a popular tool for testing RESTful APIs. It allows developers to easily create, send, and analyze HTTP requests. It has a user-friendly interface and supports various features, such as request and response validation, environment variables, and automated testing. It also allows us to test end-to-end scenarios, and it can be integrated with other tools such as Jenkins.

It's worth noting that testing RESTful APIs is an ongoing process that should be done throughout the development process, not just at the end. This will help to ensure that an API works as expected and any issues are identified and resolved quickly.

In this chapter, we have demonstrated testing an API using Thunder Client inside VS Code. This is a very useful tool, with the benefit that the tests that are defined are saved in the repo and are checked against the code.

Summary

We have covered a *lot* in this chapter! I hope it has not been overwhelming! We started at the start with a definition of REST. Then, we covered HTTP status codes and HTTP verbs to give some background on some of the underlying fundamentals of REST APIs.

We then looked at an example, covered the five most important HTTP verbs (GET, POST, DELETE, PUT, and PATCH), and showed how we could build them and test them with Thunder Client right within VS Code!

We also looked at `AutoMapper` and how to simplify object conversions to create DTOs from entity types.

Finally, we worked through an example of how to version an API and looked at some additional testing techniques.

In the next chapter, we will consider microservices, and we'll look at how we can break up this application into a number of smaller microservices!

Further reading

- HTTP response status codes: `https://developer.mozilla.org/en-US/docs/Web/HTTP/Status`

- Using HTTP Methods for RESTful Services: `https://www.restapitutorial.com/lessons/httpmethods.html`

- HATEOAS and Why It's Needed in RESTful API? `https://www.geeksforgeeks.org/hateoas-and-why-its-needed-in-restful-api/`

- Testing an API: `https://learning.postman.com/docs/designing-and-developing-your-api/testing-an-api/`

- How to use API versioning in ASP.NET Core Web API and integrate it with Swagger using .NET 6: `https://blog.christian-schou.dk/how-to-use-api-versioning-in-net-core-web-api/`

Questions

1. What are the advantages of using `PATCH` over `PUT`?

2. What is the benefit of using AutoMapper?

3. What HTTP code should be used when a new resource has been created?

4. What does the `500` range of HTTP codes signify?

5. What does REST stand for?

6. What is the purpose of DTOs?

6

Microservices for
SaaS Applications

Microservices are a powerful architectural pattern that is becoming increasingly popular in modern software development. Microservices are a way of organizing a large application into smaller, more manageable, and independent services that can communicate with each other using well-defined interfaces. This approach has gained popularity over the years as companies seek to build more scalable and robust software systems. With the increasing complexity of modern applications, it has become clear that the traditional monolithic approach to building software is no longer adequate. Microservices offer a way to break down complex applications into smaller, more manageable pieces that can be developed, deployed, and maintained independently.

Implementing microservices requires a shift in the way that we think about building and deploying software. With microservices, we move away from the monolithic approach of building everything in a single application and instead build a series of smaller, independent services that can be deployed and scaled independently. This shift in approach allows us to build more flexible and resilient systems, as we can update, test, and deploy individual services independently. At the same time, we need to ensure that our microservices can communicate with each other efficiently and that we have mechanisms in place to handle errors and failures in our distributed systems.

This chapter will explore the important considerations when building microservices, and provide a guide to help you get started with implementing microservices in your software development projects.

The following main topics will be covered in this chapter:

- Microservices and their uses
- Best practices for building performant and secure microservices
- How to use the skills we learned for RESTful APIs with microservices
- Common pitfalls when building microservices and how to avoid them
- Some practical advice

Technical requirements

All code from this chapter can be found at `https://github.com/PacktPublishing/Building-Modern-SaaS-Applications-with-C-and-.NET/tree/main/Chapter-6`.

What are microservices and why use them?

Microservices are a software architecture style that structures an application as a collection of loosely coupled services. Unlike a traditional monolithic architecture, where all the functionality of the application is bundled into a single code base, microservices break down the functionality into smaller, independent services that can be developed, deployed, and scaled independently.

The microservices architecture was first introduced in the early 2000s and has gained popularity in recent years due to the increasing demand for faster development and deployment cycles, flexible scaling, and improved application resilience. Microservices are seen as a way to deliver value to customers more quickly and with greater agility, making them ideal for organizations that are looking to rapidly iterate and innovate.

Microservices also have several benefits for organizations. By breaking down an application into smaller, focused services, organizations can increase the speed and efficiency of their development process. Teams can work on smaller, independent services in parallel, reducing the risk of delays and helping ensure that each service is delivered with high quality. Additionally, microservices can be deployed and scaled independently, allowing organizations to respond more quickly to changes in demand and optimize the performance of their applications.

When learning about microservices, it's important to appreciate the underlying concepts of loose and tight coupling.

The difference between loose and tight coupling

In a tightly coupled system, components are strongly dependent on one another and have a close relationship with each other. This means that changes in one component can have a significant and often breaking impact on other components in the system. Tightly coupled systems can quickly become very complex and difficult to maintain. Changes in one part of the system can cause unintended consequences elsewhere. If there is limited automated testing, this can be very difficult to detect and can result in bugs in production code.

In contrast, a loosely coupled system has components that are independent and have minimal dependencies on one another. This allows components to be developed, deployed, and maintained independently, with little or no impact on other components in the system. Loose coupling enables teams to work more efficiently and with greater agility, as changes to one component can be made with little or no impact on other parts of the system. It may sound like loose coupling is an obvious choice, but in practice, a lot of additional work is required to design a loosely coupled system.

The microservices architecture is based on the principles of loose coupling, where an application is broken down into smaller, focused services that can be developed, deployed, and scaled independently. This allows organizations to respond more quickly to changes in demand, optimize the performance of their applications, and increase the speed and efficiency of their development process.

SaaS applications are typically quite complex. There will often be quite large teams working on the project, each of which will have areas of specialization. By breaking down an application into smaller, focused services, teams can work on them in parallel, reducing the risk of delays and increasing the speed of delivery. This helps organizations respond more quickly to changes in demand and deliver value to customers more rapidly.

Another advantage of microservices is better scalability. Leaning on cloud-based infrastructure, each service can be deployed and scaled independently, allowing organizations to respond more quickly to changes in demand and optimize the performance of their applications. This is particularly important for SaaS applications, which often experience fluctuations in usage patterns. This can benefit the performance of the application, by provisioning more resources when the usage is high. When usage is low, some resources can be turned off, helping the organization manage its cloud compute costs and optimize the use of resources.

Microservices should also be designed to be highly resilient, meaning that even if one service fails, the impact on the overall system is minimal. This makes microservices ideal for SaaS applications, where downtime can have a significant impact on customer satisfaction and an according impact on the revenue of the company delivering the application. By breaking down an application into smaller, focused services, organizations can reduce the risk of unintended consequences when making changes to the system, making maintenance and modification easier and reducing the risk of downtime.

Microservices are a valuable tool for organizations looking to develop SaaS applications. They offer increased agility, better scalability, improved resilience, easier maintenance, and better cost management, making them an ideal choice for organizations looking to rapidly iterate and innovate.

Docker

With the `GoodHabits` service, we have been using Docker to run a `devcontainer` that encompasses the development environment, and the SQL server database. This is only one way that Docker can be used. At its core, Docker is a tool that can be used to run a process in a container. You can think of a container as a very lightweight virtual machine, typically running a Linux distro.

Docker often becomes a very important tool when building with microservices. A microservice app will typically have many different components that must be running for the overall system to run. This could involve running code in different programming languages, and running against multiple different database platforms. Attempting to get everything running reliably on a dev machine, and in several cloud environments, can quickly become a nightmare!

Docker provides an efficient and reliable way to run your microservices in a set of networked containers. By containerizing your microservices, you can isolate them from each other, making them easier to build and run.

Additionally, Docker allows you to easily package your microservices and their dependencies, making it simpler to deploy your services in different environments, including running them in a development environment. This helps ensure that your microservices will work consistently across different systems, which is essential, and also challenging when building microservices.

While Docker is not strictly necessary for building microservices, it is highly recommended as a best practice to improve the efficiency and reliability of your microservices deployment. We'll cover some more best practices in detail in the next section.

Best practices for building microservices

When building microservices, it's important to think about best practices to ensure that you get the benefits from the additional work required to build a microservice application. The system should be scalable, maintainable, and resilient – payback for the extra effort! Here are some of the most important best practices or "principles" to keep in mind.

Design for failure

Microservices should be "designed for failure." If they are going to fail (and all software will fail!), they should fail gracefully, with redundancy built in to ensure that the system continues to function even if one service fails.

One of the essential steps when designing for failure is *adding redundancy to your system*. This can be achieved by having multiple instances of each service running in parallel so that if one fails, others can continue to operate. Note that this will incur some additional costs, such as cloud hosting costs. Load balancing helps distribute the load evenly across multiple instances of a service, reducing the risk of a single instance becoming overwhelmed, and also helps redirect the load to another instance if one instance fails.

Circuit breakers are another useful tool when designing for failure. These can be used to automatically detect and isolate failing services, preventing them from affecting the rest of the system. This makes it more likely that the overall system will continue to operate, even if one service fails.

Idempotence is also crucial when embracing the "design for failure" principle. This involves ensuring that each service is idempotent, meaning that it can be executed multiple times with the same result. This allows you to retry requests if a service fails, reducing the risk of data loss or inconsistent results. You will remember that we encountered this topic in the previous chapter when learning about certain HTTP verbs. The principle is the same here.

Health checks should be used to periodically test each service and determine whether it is operating correctly. This information can then be used to automatically redirect requests to other instances of the service if the original instance fails (leaning on the redundancy principle). These health checks should run automatically or on a defined schedule and can alert the team immediately if any issues arise.

Focus on decoupling

Microservices should be (by definition) loosely coupled, with minimal dependencies between services. This allows services to be developed, deployed, and modified independently, reducing the risk of unintended consequences.

Decoupling is a central tenet and is an important aspect of building a microservices-based system. Decoupling refers to separating the concerns between different services, allowing them to operate independently and reducing the risk of cascading failures.

There must be *clear service boundaries* – each service should have a specific responsibility and purpose, without overlapping with other services. Doing so will help ensure that each service can be developed and deployed independently, reducing the interdependencies between services.

Asynchronous communication is an important aspect of decoupling services. Instead of direct communication, messages can be sent between services and processed at a later time. This allows each service to operate independently, reducing the risk of blocking and cascading failures.

It is very important to *implement some form of versioning* when decoupling services. Each service should have a version number, which can allow multiple versions of a service to coexist in the system. This allows for changes to be made to a service without them affecting other services, reducing the interdependencies between services. We looked at versioning in the previous chapter.

Using an event-driven architecture is another important part of decoupling services. Events can trigger actions in other services, reducing the need for direct communication. This allows each service to operate independently, reducing the interdependencies between services. This is often seen with event-based message queues facilitating communication between the various services in the system.

Finally, *service discovery* is a useful tool for decoupling services. If you consider a system that has maybe 20 different loosely coupled services hosted in the cloud in a constellation of Docker containers, and perhaps multiple versions of some of them, keeping track of where they are all running can become very challenging. Using some form of service discovery allows the system to detect and connect to other services automatically, reducing the need for hardcoded connections.

Embracing the "focus on decoupling" principle helps in building a robust microservices-based system. By defining clear service boundaries, using asynchronous communication, implementing versioning, using event-driven architecture, and considering service discovery, you can ensure that your system is scalable, flexible, and resilient.

Embrace automation

Automation is critical for the efficient operation of microservices as it helps ensure consistency and reliability across services. Automation should be used as much as possible to improve testing (with an automated test suite), deployment (CI/CD), and scaling (perhaps with Terraform).

Automation is a very important principle in all types of software development, but this is doubly true when building a SaaS application that makes use of a microservice architecture. Automating deployment and testing processes helps in reducing manual efforts and increases the speed of delivering new features to users. Automated deployment and testing processes ensure that services are deployed consistently, and any issues can be detected early in the development cycle. This helps in reducing downtime and increasing the overall efficiency of the system.

Implementing CI/CD helps ensure that code changes are automatically built, tested, and deployed. This helps in reducing the manual efforts involved in testing new features and gets them into the hands of the users as quickly as possible. CI/CD also helps in ensuring that code changes are deployed consistently and any issues are detected early in the development cycle. Using pipelines to automatically build, test, and deploy microservices will make managing the project as it starts to grow considerably easier!

It is also useful to automate monitoring and logging. Automating monitoring and logging helps in detecting issues early and reduces downtime. Automated monitoring and logging processes ensure that the system is monitored consistently and that any issues are detected early, reducing manual efforts and increasing the overall efficiency of the system.

When in production, a SaaS application can experience rapid fluctuations in the demands being placed on various parts of the system. Automation can facilitate automatic scaling to ensure that the system can handle increased traffic without manual intervention. Automated scaling processes ensure that the system can scale up or down based on the traffic, reducing manual efforts and increasing the overall efficiency of the system.

Embracing the "embrace automation" principle helps in building a robust and efficient microservices-based system. Automating deployment and testing processes, implementing CI/CD, automating monitoring and logging, and automating scaling processes help in streamlining the processes, reducing manual efforts, and increasing the efficiency of the system.

Use contract-driven development

Microservices should have well-defined contracts that define the interface between services. This allows services to evolve independently, while still ensuring compatibility. In this context, a "contract" means an agreement that specifies the interactions between services, including details about the inputs and outputs of each service, communication protocols, and data formats. This contract can be represented in various forms, such as API specifications, message formats, or documentation, and should be agreed upon by all the teams involved in building and maintaining the services.

Contract-driven development requires that clear contracts between services are defined. These contracts should define the inputs and outputs for each service and ensure that the services operate as expected. This helps in reducing the risk of breaking changes and increases the overall stability of the system.

As with many parts of the application, testing is very important. *Contract testing* ensures that the contracts between services are tested and adhered to, reducing the risk of breaking changes and increasing the overall stability of the system.

The "use contract-driven development" principle helps in building a robust and stable microservices-based system. Defining clear contracts between services, testing contracts, implementing contract testing, and automating contract testing help in ensuring that the services are operating as expected and adhering to the defined interface, reducing the risk of breaking changes and increasing the overall stability of the system.

Monitor and log aggressively

Microservices systems generate a large amount of data, and it's important to have a robust monitoring and logging strategy in place. This will help you detect and diagnose issues – hopefully, before they have an impact on your user!

The system should be continuously and automatically monitored, with the monitoring covering the overall health of the service, the response times of the whole system and each microservice, and resource utilization.

Alongside the monitoring solution, there should also be a logging mechanism. Logging helps in tracking the activities of the system, detecting issues, and troubleshooting any problems. This logging should include logging requests, response times, and any error messages.

Using centralized logging and monitoring helps in reducing manual efforts and increases the efficiency of the system. Centralized logging and monitoring ensure that the logs and the monitoring data are stored in a single place, making it easier to detect issues and troubleshoot problems.

There is no point in doing this monitoring and logging if the system never reports any issues! Automated alerting processes ensure that any issues are detected early and the appropriate team is notified, reducing manual efforts and increasing the overall efficiency of the system.

Monitoring, logging, and alerting help ensure the system is robust and efficient. While they do add a little bit of additional work to the development process, which is not seen when the system is running smoothly, they are very much worth the time invested when things inevitably go wrong!

Implement security

It does somewhat go without saying, but microservices should be secure, with appropriate authentication, authorization, and encryption protocols in place. It's also important to have a security strategy that covers the entire system, including the network, the infrastructure, and the services themselves.

The nature of microservices is such that security can be challenging. There will often be multiple containers running all sorts of different software, each with its own set of attack vectors.

Embracing the "implement security" principle is an essential aspect of building a microservices-based system. Security helps in protecting sensitive information, reducing the risk of security breaches, and ensuring the confidentiality and integrity of data. We must follow some steps to embrace this principle.

Implementing authentication and authorization is the first step toward embracing the "implement security" principle. Authentication and authorization help in ensuring that only authorized users can access sensitive information, reducing the risk of security breaches. The authentication and authorization process should be robust and secure to ensure the confidentiality and integrity of data and must encompass every part of the system.

Encrypting sensitive data is a crucial aspect of embracing the "implement security" principle. Encryption helps in protecting sensitive information, reducing the risk of security breaches, and ensuring the confidentiality and integrity of data. The encryption should be applied to all sensitive data, including data at rest and data in transit.

Because there are so many moving parts in a SaaS/microservice-based system, security should be implemented at a network level to encompass every part of the system. The security measures at the network level should include firewalls, intrusion detection and prevention systems, and network segmentation.

Security is extremely important in any application. SaaS applications typically have complex security requirements. Embracing the "implement security" principle from the very start of the project helps in building a secure and reliable microservices-based SaaS application. Implementing authentication and authorization, encrypting sensitive data, and implementing security at the network level help in reducing the risk of security breaches and ensuring the confidentiality and integrity of data.

Focus on scalability

Microservices should be designed to be scalable, both horizontally (by adding more instances) and vertically (by adding more resources to existing instances). This will allow you to respond quickly to changes in demand and ensure that the system continues to perform well under heavy load.

Embracing the "focus on scalability" principle is another important aspect of building a microservices-based SaaS application. Scalability helps in ensuring that the system can handle increased traffic, reduce downtime, and improve the overall performance of the system.

Designing for scalability is the first step toward embracing the "focus on scalability" principle. Scalable design helps in ensuring that the system can handle increased traffic, reduce downtime, and improve the overall performance of the system. The design should take into account the expected traffic and resource utilization and should include provisions for increasing the resources as needed.

As we discussed earlier, containerization helps in improving the scalability of the system by packaging the services and their dependencies, making it easier to deploy and scale the services as needed.

To get the most benefit from a containerized application, you should include a load balancer. Load balancing helps in distributing the traffic evenly across the available resources, reducing downtime and improving the overall performance of the system.

With containers and a load balancer, it is possible to automate scaling in the application. Implementing auto-scaling helps in ensuring that the system can handle increased traffic, reduce downtime, and improve the overall performance of the system. Auto-scaling automatically increases or decreases the resources as needed, based on the traffic and resource utilization.

Designing for scalability, embracing containerization, implementing load balancing, and implementing auto-scaling help in ensuring that the system can handle increased traffic, reduce downtime, and improve the overall performance of the system.

Separate data stores

Most applications have a single database that stores all of the information associated with the application. While this approach can be taken in a microservices application, you can also implement a database on a per-service basis. There are pros and cons to this approach, and you don't necessarily need to use a separate data store per service. The choice of using a separate data store per service or a shared data store depends on the requirements and constraints of your system.

Having separate data stores for each service can provide benefits such as these:

- **Improved scalability**: Each service can scale its data store independently, allowing for better resource utilization and reducing the likelihood of resource contention

- **Improved resilience**: Each service can continue to function, even if one of the data stores experiences an issue

- **Improved data isolation**: Each service has complete control over its data, making it easier to maintain data consistency and integrity

However, separate data stores can also introduce challenges:

- **Increased operational complexity**: Managing multiple data stores can be more complex than managing a single data store

- **Increased latency**: Communication between services to access data in different data stores can introduce latency

- **Increased data duplication**: The same data may need to be stored in multiple data stores, increasing storage costs and the risk of data inconsistency

A shared data store, on the other hand, can simplify the architecture and reduce operational complexity, but it can also introduce constraints on scalability, resilience, and data isolation.

Ultimately, the choice between separate data stores and a shared data store depends on the specific requirements and constraints of your system and should be made based on a careful evaluation of the trade-offs involved.

Designing microservice applications is hard. There is a lot to think about to make sure that you can realize the benefits of this approach! These best practices will help you build microservices that are scalable, maintainable, and resilient. By following them, you can ensure that your microservices-based system is optimized for performance, efficiency, and reliability!

Mixing microservices and RESTful APIs

When building a microservices-based architecture, REST is often used as the communication protocol between the different services. **REST**, or **Representational State Transfer**, is a commonly used and widely adopted web service architecture that provides a standardized way for clients and servers to communicate with each other. Microservices and REST are a natural fit, as REST provides the necessary communication infrastructure for microservices to communicate and exchange data with each other. We discussed contract-based development previously; the surface of the REST API can be seen as the contract for communication between services.

One of the key advantages of using REST in a microservices-based system is that it provides a clear and standard way for services to communicate with each other. REST defines a set of rules for how services should exchange data, including the use of HTTP methods such as GET, POST, and DELETE, and the use of HTTP status codes to indicate success or failure. This makes it easy for developers to build and maintain microservices, as they know exactly what to expect when communicating with other services.

Another advantage of using REST in a microservices-based system is that it provides a scalable and flexible way for services to communicate with each other. REST is platform-agnostic, typically communicating over HTTP, which means that it can be used with a variety of programming languages and technologies, making it an ideal choice for building microservices.

Finally, using REST in a microservices-based system provides a secure way for services to communicate with each other. REST uses standard web security measures such as SSL/TLS encryption, which helps protect data in transit, and HTTP authentication, which helps ensure that only authorized clients can access the data.

Microservices and REST are a natural fit, and using REST as the communication protocol between microservices provides a clear, scalable, and secure way for services to communicate and exchange data with each other. By using REST, developers can build and maintain microservices-based systems with confidence, knowing that they have a reliable and widely adopted communication infrastructure in place.

Splitting up a single REST API into microservices

When you think of a "typical" RESTful API, you will probably be thinking of a system with several controllers, each grouping several related methods or endpoints. It is not uncommon for an enterprise system with a single **monolithic** API to have dozens of controllers and hundreds of endpoints. Splitting this up into a contract-driven microservice-based system is not easy. There is no one correct way to approach this, and it can be more art than science.

Here are a few approaches that can be taken for splitting up a monolithic REST API into microservices:

- **Functionality-based**: This approach involves breaking down the monolithic API into smaller services based on the functionality they provide. For example, a service could be created to handle user authentication, while another could handle product management. This approach makes it easier to manage and maintain the services, as each one is focused on a specific task.

- **Data-driven**: In this approach, the monolithic API is broken down into services based on the data they manage. For example, a service could be created to manage customer information, while another could manage order information. This approach is useful when there are different data access patterns, security requirements, or performance requirements for different datasets.

- **Domain-driven**: This approach involves breaking down the monolithic API into services based on the domain it represents. For example, a service could be created to manage information about customers, while another could manage information about products. This approach is useful when there are complex business domains that can be broken down into smaller, manageable pieces.

- **Micro frontends**: This approach involves breaking down the monolithic API into microservices and using a micro frontend architecture to combine the services into a single user interface. This approach provides a way to scale the frontend and the backend independently, while still providing a seamless user experience.

Regardless of the approach used, it is important to consider the complexity of the API, the dependencies between the different parts of the API, and the skills and resources of the development team when determining the best way to split up a monolithic API into microservices. Additionally, it is important to continuously evaluate and refactor the microservices as needed to ensure that they continue to meet the needs of the application and the business.

An often-asked question when discussing REST and microservices is, *"Should each controller in an API be its own microservice?"*

The answer is not straightforward and depends on the specific requirements of your system and the size and complexity of each controller. In general, each microservice should represent a single, self-contained business capability, and multiple controllers can be part of a single microservice if they work together to provide a single business function.

If the controllers are tightly coupled and cannot be separated, it might make sense to have them in a single microservice. On the other hand, if each controller has separate business logic and data storage, and can be deployed and scaled independently, it might be a better fit to have each controller in its own microservice.

The key is to determine the business functions that need to be performed and to decompose the system into a set of self-contained microservices that can be developed, deployed, and scaled independently. When in doubt, it is better to start with smaller microservices and consolidate them later if needed. This allows for easier testing and debugging, as well as more rapid development and deployment cycles.

Each controller in an API doesn't necessarily have to be its own microservice, but the decision should be based on the specific requirements of your system and the size and complexity of each controller.

When combining microservices and REST, there are several important topics to cover to build a robust and scalable system, which play a crucial role in building a microservices-based system:

- **Designing RESTful APIs**: RESTful APIs should be designed to be scalable, flexible, and easy to consume.

- **API documentation**: API documentation should be clear, concise, and easy to understand, and should provide clear instructions on how to consume the APIs.

- **API versioning**: API versioning helps in ensuring that the system can evolve, without breaking existing integrations.

- **API security**: API security should be implemented to protect against unauthorized access, data theft, and other security risks.

- **Error handling**: Error handling should be implemented to ensure that the system can handle and respond to errors consistently and predictably.

- **Data consistency**: Data consistency is a crucial aspect of building a microservices-based system. Data consistency should be maintained across the microservices, to ensure that the system behaves as expected.

When combining microservices and REST, it is important to focus on designing RESTful APIs, providing clear API documentation, implementing API versioning, securing the APIs, handling errors, and maintaining data consistency. These topics help in building a robust and scalable system that can handle increased traffic and provide a better user experience.

Common pitfalls and how to avoid them

Building SaaS apps is hard. Building a microservice application is hard. Combining the two is really hard, and there are several common pitfalls that you should avoid!

The first, most common, and most important one to avoid is building the microservices too early. It is often easier to start with a monolith and slowly break off small sections of the app into small, self-contained services when the need arises and not before.

So, perhaps the best advice I can give about microservices is to not use them... until you have to! But, given that this is a chapter about microservices, here are some common traps that should be avoided if you have decided to go down this route!

- **Over-complication**: One of the most common pitfalls when building microservices is over-complicating the architecture. This can result in additional bugs, increased maintenance costs, and longer development times.

- **Lack of communication and coordination**: When building microservices, it is important to ensure that there is effective communication and coordination between teams. Without this, there can be delays and misunderstandings, which can result in problems with the overall architecture, which will inevitably manifest as a degraded experience for the users of the system.

- **Inconsistent data**: When using microservices, it is important to ensure that data is consistent across all services. Otherwise, it can lead to problems with data integrity and accuracy.

- **Increased deployment complexity**: Microservices can increase deployment complexity as each service must be deployed and managed individually.

- **Debugging complexity**: Debugging complex microservice architectures can be more difficult than debugging a monolithic architecture.

Here's how to avoid the common pitfalls when building microservices:

- **Over-complicated microservice architecture**: This can be avoided by keeping the architecture simple, focusing on the single responsibility principle, and defining clear boundaries for each microservice. It is also important to prioritize the microservices that need to be built and make sure they align with business goals.

- **Lack of communication and collaboration between teams**: This can be mitigated by creating a culture of collaboration and having clear communication channels between teams. It's also important to have regular meetings between teams to ensure that everyone is on the same page.

- **Underestimating the complexity of data management**: To avoid this, it's important to properly plan out the data management strategy for each microservice. This includes defining the data sources, data ownership, and data consistency. Using a data management solution, such as a data management platform or event sourcing, can also help.

- **Not adequately monitoring microservices**: To avoid this pitfall, it's important to have a solid monitoring strategy in place. This includes setting up logging and monitoring for each microservice and having alerting mechanisms in place for when things go wrong.

- **Lack of security considerations**: To avoid this, it's important to have a solid security strategy in place. This includes considering security at every stage of the microservice development process, including architecture, design, development, and deployment. It's also important to regularly review the security strategy and make changes as needed.

Some practical advice

Microservices are a huge and complex topic – rather than trying to give a demonstration of a full microservice application here, I will instead offer some practical advice using the demo app we have been building as the basis for this advice. The implementation is left up to you!

A microservice architecture example

It is worth reiterating that in many cases, the best approach when planning a new application is to start with a single monolith app and carve off sections of the app into microservices when the need arises.

For this example, I will assume that the Good Habits app has grown to the point where it is necessary to start thinking about separating it into microservices. I think that a useful way to split this up might be something like this:

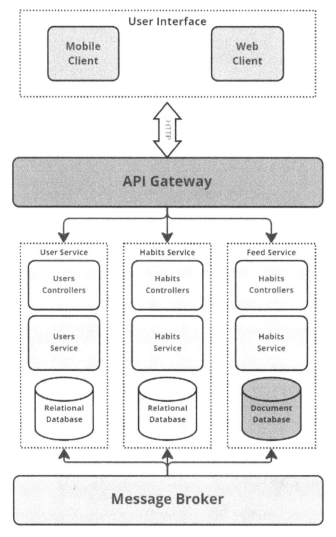

Figure 6.1 – Suggested microservice architecture

Let's discuss the components in this diagram a bit more.

User interfaces and clients

It is very common these days for an application to have both a web client app and a mobile app – and sometimes, a desktop application as well. All of these clients will be communicating with the same backend and the same microservices. It is common to build a single **gateway** API that manages all external communication between the various microservices.

API gateway

Although it's possible to enable user interface applications to directly communicate with microservices, managing this can become incredibly complex. The diagram only displays three microservices, but there could easily be 20 or more. Consider the added complexity of orchestrating communication between more than 20 microservices and three or more types of client UI – the situation becomes even more difficult to visualize and much harder to manage in practice!

Using an API gateway in a microservice architecture provides several benefits. Firstly, an API gateway acts as a single entry point for clients, allowing for easier management of requests, authentication, and authorization. It also enables the composition of different microservices into a unified API, which can simplify the client's interaction with the system. The API gateway can also provide load balancing and failover capabilities, which are important for high-availability systems. Another important benefit is the ability to enforce security and traffic policies, as well as to monitor and log requests. By using an API gateway, developers can more easily manage and evolve a microservice architecture, while maintaining a high level of security and performance.

Message broker

A message broker is used to facilitate communication between the various backend microservices. This performs much the same function on the backend as the API gateway does for the frontend. It detangles all of the communication between the services. Again, while we only have three services in the diagram, we should keep in mind that a real-world system may have many more, and inter-service communication can quickly become extremely complex and inefficient.

Using a message broker in a microservice architecture provides many benefits. One of the main advantages is that it allows services to communicate with each other asynchronously, decoupling the sender from the receiver. This can improve reliability and scalability as services can process messages at their own pace and are not blocked by the performance of other services. A message broker can also act as a buffer between services, which can be particularly useful if one service is temporarily unavailable. It can help avoid message loss by persisting messages until they can be delivered to the appropriate service. A message broker can also provide a centralized way to monitor and manage the flow of messages between services, making it easier to track and debug issues. Finally, by separating the communication concerns from the business logic, services can be more easily tested and deployed independently.

Using a message broker (and an API gateway) leans into a number of the microservice good design principles that we discussed earlier in this chapter.

Several message brokers are commonly used in .NET microservice applications, including the following:

- **RabbitMQ**: An open source message broker that supports multiple messaging protocols, including AMQP, MQTT, and STOMP

- **Apache Kafka**: A distributed streaming platform that is optimized for handling high-volume and high-velocity data streams

- **Azure Service Bus**: A fully-managed messaging service provided by Microsoft Azure that supports both traditional messaging patterns and pub/sub scenarios

- **AWS Simple Queue Service (SQS)**: A fully-managed message queue service provided by Amazon Web Services that decouples and scales microservices, distributed systems, and serverless applications

- **NServiceBus**: A messaging framework for .NET that provides a unified programming model for building distributed systems using a variety of messaging patterns

All of these tools provide reliable message delivery, scalability, and fault tolerance, and they can help simplify the communication between microservices in a distributed system.

The services

To give an example, I have laid out three services:

- A user service, which will handle everything related to the users. This includes authentication, password management, and keeping personal information up to date.

- A habit service, which deals with everything related to the habits that the users are trying to track.

- A feed service. In this more advanced version of the Good Habits app, I have assumed that there will be the ability to see your friends' progress in a social media-style feed.

Please note that I have chosen to also separate the data stores into individual databases. This allows us to treat each data store slightly differently. I have also decided to use a relational database for the User and the Habit service, but a document (NoSQL) database for the Feed service. This is a superpower of microservices – you can use different types of data storage based on the use case for individual microservices.

The User service

I have specifically broken this out because storing users' private data should be taken very seriously. This data may include banking information (if the service has a paid tier) and may include personal information. It is easy to understand that the data that is stored in the User service may have to be treated more securely than the data in the Habit service.

We should also consider that users in many jurisdictions have the right to be forgotten. By grouping all of the personal data in one place, we make this easier to achieve.

The User service would probably not be subject to particularly high demand. Users would be expected to update their details infrequently, so this service could have fewer resources allocated to it.

The Habit service

In the Good Habits application, it would be expected that the Habit service would do most of the heavy lifting in the application, and so would have additional resources allocated to it. This service should also be designed to scale easily so that more development time can be put into the performance, and perhaps less attention is paid to security than for the User service. (Of course, security is still important!!)

The type of data in this service would be highly relational, so a relational store is the most appropriate.

The Feed service

If we envisage a much more advanced version of the Good Habits app, we may have expanded to the point where we have a social network-type sets of features, which allows users to see their friends' progress and celebrate their success with them.

This type of service is usually modeled using a document store, or something like GraphQL. A relational store is not appropriate. Using a microservice architecture allows us to choose the most appropriate type of data store!

All of the information in the feed will have been selected by the users to make publicly available, so data security is less important in this service. We can be sure that there is no private data in this service, as the user data stores are not accessible from this service.

Overall architecture

The architecture in *Figure 6.1* shows one way that we could use microservices to break up the application. But there are many possible ways to do this, and it is more art than science.

The most important thing is to be guided by your users, and mindful of how the choices you make in splitting up the monolith will impact the users.

A practical example

While it is outside the scope of this chapter to provide a complete example of a microservices project, we can build out some of the principles in the GoodHabits project to cement our understanding of the preceding advice.

To illustrate this, we can do the following:

1. Add a very basic `UserService`, to show how we can interact with multiple microservices.

2. Add an API gateway that serves as a single point of entry to all of the microservices from any clients interacting with the system.

UserService

Run the following script to add the project and files for the User service:

```
dotnet new webapi -n GoodHabits.UserService; \
cd GoodHabits.UserService; \
rm ./WeatherForecast.cs; \
rm ./Controllers/WeatherForecastController.cs; \
touch Controllers/UsersController.cs; \
dotnet add package Microsoft.AspNetCore.Mvc.Versioning; \
dotnet add reference ../GoodHabits.Database/GoodHabits.Database.
csproj; \
cd ..; \
dotnet sln add ./GoodHabits.UserService/GoodHabits.UserService.csproj;
```

Next, we will configure how to start the Users microservices. Set up the `launchSettings.json` file so that it looks like this:

```
{
  "$schema": "https://json.schemastore.org/launchsettings.json",
  "profiles": {
    "UserService": {
      "commandName": "Project",
      "dotnetRunMessages": true,
      "launchBrowser": false,
      "applicationUrl": "http://localhost:5200",
      "environmentVariables": {
        "ASPNETCORE_ENVIRONMENT": "Development"
      }
    }
  }
}
```

Finally, add the following code to the controller:

```
using GoodHabits.Database.Entities;
using Microsoft.AspNetCore.Mvc;

namespace GoodHabits.UserService.Controllers;
```

```
[Route("api/[controller]")]
public class UsersController : ControllerBase
{
    private readonly Ilogger<UsersController> _logger;

    public UsersController(
        Ilogger<UsersController> logger
        )
    {
        _logger = logger;
    }

    [HttpGet()]
    public async Task<IactionResult> GetAsync()
    {
        return Ok(new List<User>()
        {
            new User() { Id = 111, FirstName = "Roger",
              LastName = "Waters", Email = "rw@pf.com"},
            new User() { Id = 222, FirstName = "Dave",
              LastName = "Gilmore", Email = "dg@pf.com"},
            new User() { Id = 333, FirstName = "Nick",
              LastName = "Mason", Email = "nm@pf.com"}
        });
    }
}
```

That is all that is required to set up a very simple User service. You can start this up independently and see how it works with Swagger. The provided functionality is very basic, and it would be a fantastic exercise to try to build this service out a bit more.

API gateway

As described previously, an API gateway gives a client using the application a single point of entry into the application. All they need to do is talk to the gateway, and the complexity of the microservice implementation is kept hidden.

We will use a package called Ocelot, which provides most of the functionality that we will need out of the box. To get started, execute the following script to set up the ApiGateway project:

```
dotnet new webapi -n GoodHabits.ApiGateway; \
cd GoodHabits.ApiGateway; \
rm ./WeatherForecast.cs; \
rm ./Controllers/WeatherForecastController.cs; \
dotnet add package Ocelot; \
```

```
dotnet add package Ocelot.Cache.CacheManager; \
dotnet sln add ./GoodHabits.ApiGateway/GoodHabits.ApiGateway.csproj; \
touch ocelot.json;
```

As we did with `UserService`, we will need to modify the `launchsettings.json` file to configure how the API gateway starts. Set up the file like this:

```
{
  "$schema": "https://json.schemastore.org/launchsettings.json",
  "profiles": {
    "ApiGateway": {
      "commandName": "Project",
      "launchBrowser": true,
      "launchUrl": "swagger",
      "environmentVariables": {
        "ASPNETCORE_ENVIRONMENT": "Development"
      },
      "applicationUrl": "http://localhost:5300",
      "dotnetRunMessages": true
    }
  }
}
```

Next, make the `Program.cs` file look like this:

```
using Ocelot.Cache.CacheManager;
using Ocelot.DependencyInjection;
using Ocelot.Middleware;

var builder = WebApplication.CreateBuilder(args);

builder.Services.AddControllers();
builder.Services.AddEndpointsApiExplorer();
builder.Services.AddSwaggerGen();

builder.Configuration.AddJsonFile("ocelot.json", optional: false,
reloadOnChange: true);
builder.Services.AddOcelot(builder.Configuration)
    .AddCacheManager(x =>
    {
        x.WithDictionaryHandle();
    });

var app = builder.Build();
```

```
if (app.Environment.IsDevelopment())
{
    app.UseSwagger();
    app.UseSwaggerUI();
    app.UseCors(policy =>
        policy.AllowAnyOrigin()
                .AllowAnyHeader()
                .AllowAnyMethod()
                );
}

app.UseHttpsRedirection();
app.UseAuthorization();
app.MapControllers();

await app.UseOcelot();

app.Run();
```

Here, you can see the key lines for the Ocelot package.

Finally, configure Ocelot by adding the following config to Ocelot.json:

```
{
    "GlobalConfiguration": {
      "BaseUrl": "http://localhost:5900"
    },
    "Routes": [
      {
        "UpstreamPathTemplate": "/gateway/habits",
        "UpstreamHttpMethod": [ "Get", "Post" ],
        "DownstreamPathTemplate": "/api/habits",
        "DownstreamScheme": "http",
        "DownstreamHostAndPorts": [
          {
            "Host": "localhost",
            "Port": 5100
          }
        ],
        "RateLimitOptions": {
          "EnableRateLimiting": true,
          "Period": "10s",
          "PeriodTimespan": 10,
          "Limit": 3
        }
      },
```

```
    {
      "UpstreamPathTemplate": "/gateway/habits/{id}",
      "UpstreamHttpMethod": [ "Get", "Delete", "Put",
        "Patch" ],
      "DownstreamPathTemplate": "/api/habits/{id}",
      "DownstreamScheme": "http",
      "DownstreamHostAndPorts": [
        {
          "Host": "localhost",
          "Port": 5100
        }
      ],
      "RateLimitOptions": {
        "EnableRateLimiting": true,
        "Period": "10s",
        "PeriodTimespan": 10,
        "Limit": 1
      }
    },
    {
      "UpstreamPathTemplate": "/gateway/users",
      "UpstreamHttpMethod": [ "Get" ],
      "DownstreamPathTemplate": "/api/users",
      "DownstreamScheme": "http",
      "DownstreamHostAndPorts": [
        {
          "Host": "localhost",
          "Port": 5200
        }
      ]
    }
  ]
}
```

If you look through the config file, you will see that we are simply mapping from one URL in the gateway to another URL in the two microservices that we have created, HabitService and UserService. This may seem like a needless complication, but if you consider that many more microservices might be added to the overall application, it makes sense to give a single point of entry.

Running the combined application

To run the complete application, we need to start all four projects (HabitService, UserService, APIGateway, and Client) individually. This can become challenging, so we will set up tasks and launch configurations to manage this for us.

In the `.vscode` folder, add the following code to `tasks.json`:

```json
{
    "label": "build-user-service",
    "type": "shell",
    "command": "dotnet",
    "args": [
        "build",
        "${workspaceFolder}/GoodHabits.UserService/
          GoodHabits.UserService.csproj"
    ],
    "group": {
        "kind": "build",
        "isDefault": true
    }
},
{
    "label": "build-api-gateway",
    "type": "shell",
    "command": "dotnet",
    "args": [
        "build",
        "${workspaceFolder}/GoodHabits.ApiGateway/
          GoodHabits.ApiGateway.csproj"
    ],
    "group": {
        "kind": "build",
        "isDefault": true
    }
}
```

In the same folder, add the following code to `launch.json`:

```json
{
    "name": "RunUserService",
    "type": "coreclr",
    "request": "launch",
    "preLaunchTask": "build-user-service",
    "program": "${workspaceFolder}/
      GoodHabits.UserService/bin/Debug/net7.0/
      GoodHabits.UserService.dll",
    "args": [],
    "cwd": "${workspaceFolder}/
      GoodHabits.UserService",
    "stopAtEntry": false,
    "console": "integratedTerminal"
```

```
            },
            {
                "name": "RunApiGateway",
                "type": "coreclr",
                "request": "launch",
                "preLaunchTask": "build-api-gateway",
                "program": "${workspaceFolder}/
                  GoodHabits.ApiGateway/bin/Debug/net7.0/
                  GoodHabits.ApiGateway.dll",
                "args": [],
                "cwd": "${workspaceFolder}/
                  GoodHabits.ApiGateway",
                "stopAtEntry": false,
                "console": "integratedTerminal"
            }
```

Also in `launch.json`, add the following compound tasks:

```
    "compounds": [
        {
            "name": "Run Server",
            "configurations": [
                "RunHabitService",
                "RunUserService",
                "RunApiGateway"
            ]
        },
        {
            "name": "Run All",
            "configurations": [
                "RunHabitService",
                "RunClient",
                "RunUserService",
                "RunApiGateway"
            ]
        }
    ]
```

The preceding configuration will allow VSCode to start all four projects by hitting the *F5* key, or by using the build and run menu.

There is much more that we could do at this point with the demo application. Some suggestions are as follows:

- Build out the User service so that it includes a lot more functionality that would be required for a real app

- Add additional routes to the `Ocelot` config

- Add a message queue (hint – try RabbitMQ)

I hope that we have managed to illustrate most of the key learnings from this chapter and provided a foundation for you to build upon.

Summary

Microservices are a huge and complex topic, much too large to tackle in one chapter of a SaaS book! In this chapter, we provided a brief introduction to microservices, covering what they are and why they are important. We discussed the benefits of using microservices, including improved scalability, fault tolerance, and flexibility. We also talked about the challenges and pitfalls of implementing a microservices architecture, such as increased complexity.

Next, we explored the common patterns for designing microservices, including service discovery, API gateways, and message brokers. We also looked at the role of containers and container orchestration systems, such as Docker, in deploying and managing microservices. Finally, we provided a set of pointers on how to implement a microservices architecture using C#, .NET, and various supporting tools. While this is only a small glimpse into the world of microservices, we hope that it has helped provide a foundation for your further exploration of this important topic.

In the next chapter, we will learn how to build a user interface with Blazor. We'll interface this UI with the Good Habits backend that we have been building in previous chapters!

Further reading

To learn more about the topics that were covered in this chapter, take a look at the following resources:

- How to build .NET Core microservices: `https://www.altkomsoftware.com/blog/microservices-service-discovery-eureka/`

- Creating a simple data-driven CRUD microservice: `https://learn.microsoft.com/en-us/dotnet/architecture/microservices/multi-container-microservice-net-applications/data-driven-crud-microservice`

- 8 Ways to Secure Your Microservices Architecture: `https://www.okta.com/resources/whitepaper/8-ways-to-secure-your-microservices-architecture/`

- Follow 6 key steps to deploy microservices in production: `https://www.techtarget.com/searchitoperations/tip/Follow-these-6-steps-to-deploy-microservices-in-production`

- Microservices with .NET: `https://dotnet.microsoft.com/en-us/apps/aspnet/microservices`

Questions

Answer the following questions to test your knowledge of this chapter:

1. What is the difference between a monolithic and microservices architecture?

2. What is the purpose of an API gateway in microservices architecture?

3. How does a message broker facilitate communication between microservices?

4. How does using a microservices architecture impact database design and management?

5. What are some common challenges or pitfalls to consider when implementing microservices?

Part 3: Building the Frontend

After learning about the backend in the previous section, we will move on to the frontend. In this section, we will build a simple **user interface** (**UI**) using Blazor and connect this to the backend from the previous section. As well as the practical skills, this section will also introduce a lot of the theory around frontend development and building excellent UIs.

This section has the following chapters:

- *Chapter 7, Building a User Interface*
- *Chapter 8, Authentication and Authorization*

7

Building a User Interface

In this chapter, we will be exploring how to build a web-based frontend using Blazor, a client-side web framework that interfaces with the .NET web API backend that we built up in previous chapters. Blazor is a powerful and flexible framework that allows us to write C# code that runs in the browser, enabling us to create rich and interactive web applications with a single code base.

We will start by looking at the many important techniques that you will have to understand before building a user interface. These will be tasks such as generating user personas, building user journeys, wireframing, and creating tickets.

After we have covered the background work, we will start by modifying the Dockerized development environment to facilitate frontend development and creating a new Blazor project. We will also explore the basic structure of a Blazor application, including the components, pages, and layouts.

Next, we will cover how to connect to a .NET web API backend. We will show how to consume data from the web API, using the client code to call the API endpoints and retrieve the data, and display it in the user interface.

We will then dive into the details of designing the user interface, including creating the layout, designing the components, and adding the necessary controls and elements to create a responsive and visually appealing user interface.

By the end of this chapter, you will have a comprehensive understanding of how to build a Blazor frontend that connects to a .NET web API backend, and how to create a responsive and scalable web application that provides a seamless UX for your customers.

The main topics covered in this chapter are as follows:

- A general introduction to the tech stack
- How to make sure your app serves your customers
- Some practical examples of how to build a UI with Blazor
- How to interact with the backend

Technical requirements

All code from this chapter can be found at `https://github.com/PacktPublishing/Building-Modern-SaaS-Applications-with-C-and-.NET/tree/main/Chapter-7`.

Introduction to the tech stack

There are many options available to build a frontend for a SaaS application. Angular and React are popular JavaScript-based frameworks that are commonly used and are as good a choice as any to build a frontend for a SaaS application. However, as this is a .NET-focused book, we will stick to .NET technology and use Blazor.

Blazor is a modern web application framework developed by Microsoft that allows developers to build client-side web applications, using C# and .NET instead of JavaScript. Blazor was first introduced as an experimental project in 2018 and was later released as part of .NET Core 3.0 in 2019. The main goal of Blazor is to enable developers to write full-stack web applications entirely in C# and .NET, providing a more familiar and productive development experience.

Blazor is designed to solve the problem of developing and maintaining complex, data-driven web applications that require a lot of client-side interactivity, such as a SaaS application. With Blazor, developers can write code that runs in a browser using WebAssembly, enabling them to create rich and interactive user interfaces without relying on JavaScript. Blazor also provides a wide range of built-in features and tools, such as routing, data binding, and form validation, which can help developers to build complex web applications more quickly and easily.

Blazor has become popular in recent years due to its ease of use, flexibility, and productivity. By allowing developers to use C# and .NET on the client side, Blazor provides a more consistent and familiar development experience, reducing the learning curve and enabling developers to be more productive. Additionally, Blazor's tight integration with .NET provides a seamless development experience, with built-in debugging, testing, and deployment tools.

Blazor is particularly suitable to build SaaS applications because it provides a scalable and reliable development platform that can handle a large number of users and data. Blazor's ability to interface with .NET web API backends makes it a powerful tool to create a complete end-to-end solution, with a robust backend and a responsive frontend. Additionally, Blazor's built-in features and tools to handle user input, data validation, and security make it an ideal choice to build complex and secure SaaS applications.

What is WebAssembly?

WebAssembly is a low-level binary format that enables the execution of code in a web browser. It is a portable, stack-based virtual machine that runs alongside JavaScript and provides a faster and more efficient way to execute code in the browser. WebAssembly is designed to work in conjunction with HTML, CSS, and JavaScript, allowing developers to write web applications using a variety of programming languages, including C, C++, Rust, and AssemblyScript.

WebAssembly is different from JavaScript in that it is a compiled language, whereas JavaScript is an interpreted language. This means that code written in WebAssembly can be precompiled, making it faster to load and execute in the browser. Additionally, WebAssembly provides a more secure and sandboxed execution environment, which can help prevent security vulnerabilities and improve the reliability of web applications.

WebAssembly is becoming increasingly popular as a way to build web applications, especially for performance-intensive tasks such as gaming, image and video processing, and scientific simulations. With the increasing popularity of tools and frameworks such as Blazor, WebAssembly is also used more frequently in the context of building client-side web applications using languages other than JavaScript. Overall, WebAssembly is an exciting and powerful technology that changes the way we think about building web applications and enables a new era of innovation on the web.

The development environment

So far, we have done all of our work in Visual Studio Code. We have added extensions that allow us to manipulate .NET projects, host and interact with databases, and also execute tests on an API.

I selected Visual Studio Code for the examples in this book because it is free, cross-platform, and allows us to do all sorts of things without requiring lots of tools to be installed. Furthermore, I have used dev containers to try to ensure that everything will "just work," regardless of what computer you are running on.

I think this is really cool and also very practical when putting together a demo app in a book. I will continue this philosophy for this chapter on building a UI, but I need to acknowledge at this point that the development environment in the "main" Visual Studio application is superior to Blazor. I hope this will change in the near future and that this chapter will stand the test of time.

For the purposes of the simple example that I present here, Visual Studio Code is fine. If you are building a more complex real-world project, then you may want to invest in Visual Studio.

UX considerations – knowing your customer

User experience (UX) is of paramount importance in modern application development. How a user interacts with your SaaS application is crucial, as poor UX can result in user frustration and abandonment. Therefore, considering the user's experience should be the most important part of the process when designing a UI.

At its core, UX is the practice of designing digital products, services, and systems that are useful, usable, and enjoyable for the people who use them. In today's digital age, where people interact with technology on a daily basis, UX has become increasingly important to create successful products and services. Good UX can enhance a user's satisfaction, engagement, and loyalty, while bad UX can result in frustration, confusion, and abandonment. UX design involves understanding user needs, goals, and behaviors, and using that knowledge to inform the design of UIs and interactions. It encompasses a range of disciplines, including visual design, interaction design, information architecture, and user research, among others. As digital products and services continue to play an increasingly important role in our daily lives, UX will continue to be a critical factor in their success.

UX is important, but it is also a complex and multifaceted field that can be challenging to master. While there are some scientific methods and principles that can be applied to UX design, such as user research, usability testing, and data analysis, the process of creating a great UX often involves a significant amount of artistry and creativity. Unlike some other fields, there are no absolute right or wrong answers in UX design, and what works for one user or product may not work for another. UX designers must balance a wide range of considerations, including user needs, business goals, technical constraints, and visual aesthetics, among others. This requires a combination of analytical skills, design skills, and intuition to create a UX that is both effective and aesthetically pleasing. In short, while there are scientific methods and principles that can help guide UX design, there is still a significant amount of artistry involved in creating great UXs.

In the context of SaaS applications, UX is even more important than it is in other types of software. SaaS applications are typically subscription-based, meaning that users pay for access to the software on an ongoing basis. This means that they have the option to switch to a competitor at any time if they are not satisfied with the service they are receiving. In other words, SaaS companies are in a constant battle to retain their users, and poor UX can be the deciding factor in whether a user stays or goes. Additionally, SaaS applications are often complex and feature-rich, with a wide range of options and settings. This can make the UX even more challenging to design, as users need to be able to find what they are looking for quickly and easily. As a result, designing a great UX is crucial to the success of a SaaS application, as it can help to increase user satisfaction, reduce churn, and ultimately drive the success of the business.

In this section, I'll give some general pointers that may be useful to consider when you build a SaaS application.

User personas

User personas are fictional characters that represent different types of users who may interact with an application. They are developed by gathering information about real users through surveys, interviews, and other forms of research. UX designers use this information to create a set of user personas that represent the different needs, behaviors, and motivations of their target users. User personas are important in the UX design process because they help to create a clear understanding of the people who will be using the application. By understanding the characteristics of the different personas, designers can make informed decisions about how to structure the UI, what features to include, and how to prioritize different aspects of the UX. For example, if the target audience includes some individuals who are tech-savvy individuals and some less technically proficient, the designer may need to create a UI that is both intuitive and easy to navigate. By creating user personas, UX designers can ensure that the design of the application is centered around the needs and expectations of its users, which can ultimately lead to a better UX and greater user satisfaction.

Here is an example of a possible fictional tech-savvy user called "Sara":

Sara is a tech-savvy user who uses multiple devices and is comfortable with technology. She has a good understanding of technology trends and new applications and enjoys exploring new features and settings in an application. Sara prefers to use the latest technology and applications and may be interested in using keyboard shortcuts and other power-user features in an application. She is comfortable with troubleshooting issues and finding solutions independently, and may have a lower tolerance for slow load times or other technical issues. If an application does not meet her expectations, she may be more likely to abandon it. Overall, Sara is a user who is comfortable with technology and has high expectations for the applications she uses, and the design of the application should reflect her needs and expectations.

Creating an avatar for a user persona is useful for designers and developers because it helps to create empathy and a better understanding of the user's needs, goals, and behaviors. We have created the following avatar for Sara!

Figure 7.1 – An avatar for Sara

User journey mapping

User journey mapping is a visual representation of the steps a user takes to complete a task within an application, from the initial point of contact through to the completion of the task. User journey mapping is important because it helps to identify pain points and areas of frustration in the UX, as well as opportunities for improvement. By mapping the user journey, designers can get a clear picture of how users interact with the application and where they may encounter issues. This information can then be used to refine the design and make the UX more streamlined and intuitive. User journey mapping is related to user personas because it helps to create a more detailed understanding of how different personas interact with the application. By mapping the user journey for different personas, designers can identify how the UX differs between different types of users and make design decisions that are tailored to the needs of each persona. Ultimately, user journey mapping is a valuable tool to create a user-centered design that meets the needs of all users.

Here's an example of a user journey for Sara, a tech-savvy user of the `GoodHabit` database:

1.   Sara navigates to the `GoodHabits` app and logs in.
2.   The app displays a page with a list of Sara's existing habits on the left side of the screen and an empty add/edit form on the right side of the screen.
3.   Sara clicks on the **Add New** button under the list of habits.
4.   The app populates the form with default values and changes the button text to **Save New**.
5.   Sara fills out the form and clicks on the **Save New** button.
6.   The app validates the form data and creates the new habit, associating it with Sara's user account.

7. The app adds the new habit to the list on the left side of the screen, and the habit is automatically selected.

8. The app updates the form on the right side of the screen with the new habit's data and changes the button text to **Save Changes**.

9. Sara can make further changes to the habit if desired, and then she clicks on the **Save Changes** button.

10. The app validates the form data and updates the habit with the new data.

11. The app highlights the updated habit on the list on the left side of the screen to indicate that it has been modified.

12. Sara verifies that the new habit has been added or updated successfully and is displayed correctly on the list.

Writing out these user journeys can be very useful for the design team, and also for the dev team, to better understand how users will interact with a system. A good understanding of these interactions will result in a better experience for the users, which will generally lead to the SaaS app generating more revenue!

It should not be assumed that everyone who uses your SaaS app will have the same level of abilities, and it is also very important to consider accessibility.

Accessibility

Accessibility refers to the practice of designing digital products and content in a way that ensures that they are usable by people with disabilities. This includes people with visual, auditory, motor, and cognitive impairments, among others. Accessibility is important because it helps to ensure that all users, regardless of their abilities, can access and use digital products and content. This not only benefits individuals with disabilities but also has broader social and economic benefits, as it helps to create a more inclusive society.

Here are some tips and things to consider when designing for accessibility:

- Providing alternative text for images so that users with visual impairments can understand the content of an image

- Ensuring that the color contrast between text and the background is sufficient so that users with visual impairments can read the text easily

- Providing closed captions or transcripts for videos and audio content so that users with auditory impairments can understand the content

- Using semantic HTML to ensure that assistive technologies can accurately parse and interpret the content of a web page

- Ensuring that an application is operable using keyboard-only navigation so that users with motor impairments can use an application easily

- Providing clear and concise instructions and feedback so that users with cognitive impairments can understand and use an application effectively

It may be helpful to create some user personas to represent users who will benefit from these accessibility considerations, building some user journeys specifically for them.

Designing for accessibility is an important consideration for any digital product or content, helping to create a more inclusive and accessible society.

Visually appealing design

While UX always blends art and science to some extent, creating a visually appealing design leans much more on artistic creativity. It is easy to overlook this when building an application and focus on the more technical aspects. However, having a visually appealing design is an important aspect of UX and can have a significant impact on how users perceive and interact with digital products and content. A well-designed UI can make it easier for users to understand and navigate an application, and it can create a sense of trust and credibility that is important to build a strong brand. A visually appealing design should be aesthetically pleasing and engaging, while also being functional and usable. This means using a consistent color scheme, typography, and layout that is easy on the eyes and provides a clear visual hierarchy. It also means using appropriate imagery and graphics that enhance content and support the overall UX. Ultimately, a visually appealing design should be intuitive, easy to use, and engaging, helping to create a positive UX for all users.

Navigation and information architecture

UX design should strive to make the navigation of an application simple and intuitive for users. This means designing a clear and consistent menu structure, as well as providing helpful labels and descriptions for each menu item. The navigation should be easily accessible from any page within the application and enable users to quickly and easily move between different areas of it.

In a SaaS app, users often try to accomplish a specific task or goal, and they need to be able to easily find the content and features they need to accomplish that task. Effective navigation can help users quickly and easily move between different areas of an app, without getting lost or confused.

Information architecture is the process of organizing and structuring the content within an application in a logical and meaningful way. It involves grouping related content together, creating hierarchies of information, and establishing clear relationships between different pieces of content. A well-designed information architecture can make it easier for users to find the information they need, and it can also help to provide context and meaning for the content within the application. When designing the information architecture, it is important to consider the needs and goals of the users, as well as the content that is being presented, in order to create a structure that is clear, intuitive, and effective.

Information architecture is important for a SaaS app because it can help to ensure that the content within the app is organized in a way that is intuitive and meaningful for users. This can help users to better understand and engage with the content, and it can also make it easier for them to find the information they need. By designing a clear and effective information architecture, SaaS app designers can create an experience that is both functional and enjoyable for users, helping to build brand loyalty and customer satisfaction.

Responsive design

Responsive design is a design approach that aims to create digital products that adapt and respond to different screen sizes and device types. This has become increasingly popular as more and more users access websites and applications on a wide variety of devices, including desktops, laptops, tablets, and smartphones. With responsive design, the layout, content, and functionality of a website or application are optimized for each device type, allowing users to have a consistent and seamless experience, regardless of the device they use.

To achieve responsive design, UX designers typically use a combination of flexible layouts, fluid images, and media queries, which allow the design to adapt to different screen sizes and resolutions. This means that elements of the design, such as the navigation, content, and images, will adjust and reposition themselves based on the size of the screen, in order to provide the best UX possible.

Responsive design helps to create a positive UX, as it ensures that users can access and use digital products on any device, at any time. With more and more users accessing the web and applications on mobile devices, responsive design has become a key consideration for UX designers and is essential to create a successful and accessible digital product.

When considering a SaaS application, this becomes even more important, as customers of the application will often require that they can access the app at any time, and on any device.

Feedback and user testing

Feedback and user testing are essential to create a successful UX, as they allow designers to gather insights and information from real users about the usability, functionality, and effectiveness of their designs. This feedback can be used to identify areas of the design that work well, as well as areas that need improvement, helping designers to refine and optimize the UX.

To collect feedback and user testing data, UX designers use a variety of techniques, including surveys, interviews, usability testing, and user analytics. Surveys and interviews can help designers to collect qualitative feedback about the UX, including likes, dislikes, and pain points. Usability testing, on the other hand, involves observing users as they interact with the design, providing valuable insights into how users use an application, as well as areas of the design that may be causing confusion or frustration. User analytics can also be used to gather data about user behavior, such as how often users access certain features, or where they drop off in the user journey.

Once feedback and user testing data have been collected, designers can use them to inform and guide their design decisions, making changes and optimizations based on the insights gathered from real users. By incorporating feedback and user testing into the design process, UX designers can create a more user-centric and effective UX, leading to higher engagement, satisfaction, and customer loyalty.

Building a simple UI

You may recall that way back in *Chapter 2*, we created a Blazor application. We will now start to build this out.

Earlier in this chapter, we imagined a user persona called Sara, and we described a user journey where Sara adds a new good habit to her database. In this section, we will build out that user journey!

Planning the UI

Building a good UI is a complex process that involves careful planning and thoughtful execution. Before starting to create the UI, it's essential to take the time to plan out the layout, design, and functionality of an application. This planning should encompass everything, from the UX to user interviews, wireframing, and prototyping.

The first step in planning a UI is to define the user personas and UX. Understanding a user's needs and the goals of an application is essential to create an interface that is both usable and engaging. One way to accomplish this is by conducting user interviews, which can help to identify pain points and opportunities for improvement. This feedback can then be used to shape the design and functionality of the UI.

Once the user personas and UX are defined, it's time to start thinking about the layout of the UI. Wireframing and prototyping are useful techniques to visualize the layout and design of the interface. These techniques allow designers to experiment with different ideas and make sure the application is intuitive and easy to navigate. Wireframes are basic sketches of the interface that help to establish the overall layout, while prototypes are interactive mock-ups that allow users to interact with the UI and give feedback.

In addition to wireframing and prototyping, it's important to consider the technical aspects of the application. This includes choosing the right technology stack, such as the appropriate framework and tools. For instance, Blazor is a popular framework to create UIs using C# and .NET, and it is an excellent choice to build a SaaS application.

Overall, planning the UI is a crucial step in creating a simple yet effective interface for a SaaS application. It involves considering the UX, conducting user interviews, wireframing and prototyping, and choosing the appropriate technology stack. With careful planning, it's possible to create a UI that is both functional and aesthetically pleasing, which will ultimately help to improve user engagement and satisfaction.

So that we can make progress, let's assume that I have done all of the preceding and decided that the following represents the ideal UI for the Good Habits SaaS application.

Figure 7.2 – A mock-up of the Good Habits app

The preceding screenshot represents a very basic UI that allows a user to see a list of their habits, add a new one, and edit or delete existing habits. This is basically just a simple **Create, Read, Update, Delete (CRUD)** implementation that would probably make a UX designer cry, but it will suffice for this demo!

Configuring the environment

As we will stick to Visual Studio Code to build the UI, we will add some extensions to make our lives a bit easier. As always, we'll do this by modifying `devcontainer.json` and add to the extensions array. Add the following extensions:

```
"kevin-chatham.aspnetcorerazor-html-css-class-completion",
"syncfusioninc.blazor-vscode-extensions",
"ms-dotnettools.vscode-dotnet-runtime",
"ms-dotnettools.blazorwasm-companion"
```

These extensions will make our lives much easier as we build out this UI!

You will need to exit and re-enter the dev container environment so that the changes can be applied.

Writing a ticket

In software development, a "ticket" is a term used to describe a unit of work that needs to be done on a project. Tickets are usually created to track bugs, feature requests, and any other tasks that need to be completed by the development team. They provide a way for project managers and developers to keep track of what needs to be done, who is responsible for completing a task, and the status of the work. Tickets can be as simple as a bug report or feature request, or they can be very detailed and include requirements, designs, and other documentation. The use of tickets helps to keep the development process organized and efficient, ensuring that everyone on a team is working toward the same goals.

Gherkin is a language that is often used to write specifications for software development projects, and it can be used when creating tickets. It is designed to be easy to understand by both technical and non-technical stakeholders, and it helps to ensure that everyone is on the same page when it comes to what needs to be done. Gherkin specifications are written in a specific format that is easy to read and understand, and they can be used to create automated tests that ensure that the software meets the requirements that have been specified. By using Gherkin to write tickets, developers can ensure that the work they do is directly tied to the needs of a project and that everyone on a team understands what needs to be done and why. This can help to reduce confusion and misunderstandings, leading to a more efficient and effective development process.

Gherkin is a plain-text, domain-specific language that is commonly used to assist in the writing of automated acceptance tests for software applications. It is designed to be human-readable and easy to understand by stakeholders, developers, and testers alike. Gherkin provides a simple syntax to define the behavior of an application in a structured and organized way, using a set of keywords and phrases.

One of the key benefits of using Gherkin to write UI tickets is that it helps to ensure clarity and consistency across the development team. By using a standardized format to describe the desired behavior of a UI, everyone involved in the development process can understand what is expected and how to implement it. Gherkin also provides a common language that can be used to communicate between team members with different backgrounds and expertise.

Another advantage of using Gherkin for UI tickets is that it promotes a user-centric approach to software development. By focusing on the desired user behavior and experience, Gherkin encourages developers to build applications that are intuitive, easy to use, and meet the needs of the end user. This approach can lead to more effective testing and a better overall UX.

Gherkin is a powerful tool to write UI tickets that can help to promote clarity, consistency, and user-centered design in software development. By providing a common language and structured format to describe behavior, Gherkin can help to ensure that everyone involved in the development process understands what is expected and how to achieve it. Additionally, by enabling automated testing, Gherkin can help to catch issues early and ensure a high-quality end product.

Here is an example of a ticket that could be used to describe the user journey that we defined for Sara previously. We will use this ticket to guide us in the development of the UI:

```
Feature: Add a new habit to the list

Scenario: User adds a new habit to the list

Given Sara is on the Good Habits page
And there is a button labeled "Add new"
When Sara clicks the "Add new" button
Then a form for adding a new habit should appear
And the form should have input fields for the habit name, description,
and category
And the form should have a "Save" button
And the form should have a "Cancel" button

Scenario: User saves a new habit to the list

Given Sara has filled out the form for adding a new habit
When she clicks the "Save" button
Then the new habit should be added to the list
And the list should show the habit name, description, and category
And the new habit should be displayed at the top of the list
And the new habit should be highlighted to indicate it has been added
successfully
And the form for adding a new habit should disappear
```

The preceding ticket gives us a roadmap to follow to develop the UI. Coupled with the wireframe that we have created previously, it should be easy to build this simple UI!

But before we get started building, let's take a quick look at the technology that we will use!

What is Blazor?

Blazor is a modern web framework to build interactive web UIs using C# and .NET. It allows developers to write web applications using the same familiar language and tools that are used for desktop and mobile applications, as well as for RESTful APIs. Blazor supports two models to build web applications – Blazor WebAssembly and Blazor Server.

Blazor WebAssembly is a client-side model that allows developers to build web applications entirely in C# and run them in any modern web browser, without the need for any server-side code. The application is loaded and executed directly in the browser, and any communication with the server is performed using standard web technologies such as HTTP and WebSockets – frequently using a WebAPI backend, which we used in previous chapters. Blazor WebAssembly applications can run both online and offline, making them suitable to build progressive web applications.

Blazor Server is a server-side model that allows developers to build web applications, using a similar programming model to client-side Blazor but with the server running the application code and rendering the UI in the browser. In this model, the application code runs on the server, and the UI is streamed to the client as HTML and JavaScript. This allows Blazor Server applications to have the same rich interactive features as client-side applications but with a higher degree of control over the UX.

Both models of Blazor provide developers with a powerful and modern way to build web applications, with the choice between client-side and server-side, depending on the specific requirements of a project. Blazor's use of C# and .NET makes it a compelling option for developers who are already familiar with these technologies, and the ability to share code between the web and other types of applications can lead to greater efficiency and productivity. Additionally, Blazor's support for Razor syntax and integration with Visual Studio and other development tools makes it a familiar and comfortable environment for developers to work in.

For this example, we will use Blazor WebAssembly (Wasm). This is a good choice when building a SaaS application with a RESTful API backend for a few reasons, some of which are stated here:

- Wasm runs entirely in the browser, which means that a user does not need to wait for server responses to load the application. This can result in faster load times, better UX, and reduced server load.

- Because the application runs entirely in the browser, it can be made available offline. This is particularly useful for users who are on the go and do not have access to reliable internet connections.

- Wasm allows more efficient processing of complex computations and graphics-intensive applications, which is particularly relevant for SaaS applications that may require a lot of data processing or advanced graphics.

- By offloading more of the processing to the client side, a Wasm-based SaaS application can reduce the amount of data that needs to be sent back and forth between a server and a client. This can result in reduced bandwidth requirements, faster load times, and reduced server load.

- Wasm can provide better security for SaaS applications, as it runs in a sandboxed environment, which makes it more difficult for malicious actors to exploit vulnerabilities in the application.

- Because Wasm is platform-agnostic, SaaS applications built using Wasm can be run on a variety of devices, including desktops, laptops, tablets, and smartphones. This can help to increase the reach of the application and make it more accessible to users.

Setting up the Blazor project

We already set up a Blazor project in *Chapter 2*, but I'll recap the steps here. Open a console and type in the following:

```
dotnet new blazorwasm -o GoodHabits.Client
cd GoodHabits.Client
```

Great! You can now start the client by typing `dotnet run` into the console. You'll see the following in the browser:

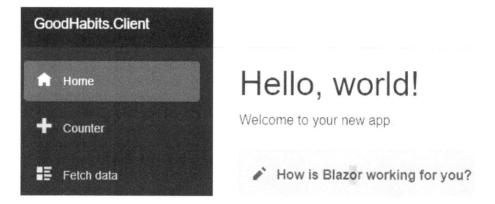

Figure 7.3 – The Hello, world! Blazor app

We'll start with this template and make it look a bit more like the mock-up. Initially, we will use dummy data, but we will connect the frontend up to the API – by the end of this chapter, we will have a full stack SaaS application!

Building the UI

First, there is a small amount of config required in `Program.cs` in the client application. Make the class look like the following:

```
using Microsoft.AspNetCore.Components.Web;
using Microsoft.AspNetCore.Components.WebAssembly.Hosting;
using GoodHabits.Client;

var builder = WebAssemblyHostBuilder.CreateDefault(args);
builder.RootComponents.Add<App>("#app");
builder.RootComponents.Add<HeadOutlet>("head::after");
var apiBaseAddress = "http://localhost:5300/gateway/";
builder.Services.AddScoped(sp => new HttpClient { BaseAddress = new
Uri(apiBaseAddress) });

await builder.Build().RunAsync();
```

Note that the preceding base address connects to the API gateway that we configured in *Chapter 6*.

With that small piece of config done, we can get started with the functionality by adding a new page that will display the information we built up in the database in the previous chapters. We will also modify the navigation menu to link to this new page, removing the sample pages.

In the Pages folder, you can delete the Counter.razor and FetchData.razor files and add a new file called Habits.razor. Your Pages folder should look like this:

Figure 7.4 – The Pages folder

With the preceding configured, we can add some basic setup on the Habits.razor page. Open the file and copy in the following:

```
@page "/goodhabits"
@inject HttpClient httpClient

<div class="row">
    <div class="col-md-4">
        <h3>Good Habits</h3>
    </div>
    <div class="col-md-8">
        <h3>Habit Details</h3>
    </div>
</div>

@code {
    private List<Habit> GoodHabits { get; set; } = new
      List<Habit>()
    {
        new Habit { Name = "Drink water", Description =
          "Drink at least 8 glasses of water every day." },
        new Habit { Name = "Exercise", Description = "Do at
          least 30 minutes of exercise every day." },
        new Habit { Name = "Meditation", Description =
          "Meditate for at least 10 minutes every day." },
    };
    private class Habit
    {
        public int Id { get; set; }
        public string Name { get; set; } = default!;
        public string Description { get; set; } = default!;
    }

}
```

The preceding code will give us an outline that we can start to build upon. However, before we can see this in the app, we need to modify the navigation menu to link to this new page.

Open the Shared folder and locate the NavMenu.razor file. Change the navigation menu code to match the following. Note that the href property matches what we set in the @page attribute on the Habits.razor page:

```
<div class="@NavMenuCssClass nav-scrollable" @onclick="ToggleNavMenu">
    <nav class="flex-column">
        <div class="nav-item px-3">
            <NavLink class="nav-link" href=""
              Match="NavLinkMatch.All">
                <span class="oi oi-home" aria-
                  hidden="true"></span> Home
            </NavLink>
        </div>
        <div class="nav-item px-3">
            <NavLink class="nav-link" href="goodhabits">
                <span class="oi oi-plus" aria-
                  hidden="true"></span> Good Habits
            </NavLink>
        </div>
    </nav>
</div>
```

With the menu configured, we can now run the application with dotnet run and see the changes we have made.

Figure 7.5 – Starting to build the GoodHabits page

In the preceding screenshot, we can see that we have successfully added the navigation and sketched out the GoodHabits page! Let's add some more functionality.

We will start by modifying the code in the code block to allow us to view, add, edit, and delete the habits. We will shortly bind these functions to UI controls so that we can manipulate the habits in the list.

In the @code block, start by adding some class-level variables:

```
private Habit? SelectedHabit { get; set; }
private Habit? EditingHabit { get; set; }
private Habit? AddingHabit { get; set; }
private bool IsEditing { get; set; } = false;
```

This will allow us to select a habit out of the list, edit that habit, and store a state variable that tells the view when we are editing.

Next, add a function that allows us to select a habit:

```
private void ShowDetails(Habit habit)
{ SelectedHabit = habit;
   IsEditing = false;
}
```

Add a function that will allow us to add a new habit with some default properties to the list:

```
private void AddHabit()
{
   Habit newHabit = new Habit()
   {
     Name = "New Habit",
     Description = "Enter a description here"
   };
   GoodHabits.Add(newHabit);
   SelectedHabit = newHabit;
}
```

Add a function that will allow us to edit a habit:

```
private void EditHabit(Habit habit)
{
   SelectedHabit = habit;
   ShowEditForm();
}
```

Add a function that will allow us to delete a habit from the list:

```
private void DeleteHabit(Habit habit)
{
   GoodHabits.Remove(habit);
   if (SelectedHabit == habit)
   {
     SelectedHabit = null;
   }
}
```

Add a function that will allow us to edit a habit from the list:

```
private void ShowEditForm()
{
  IsEditing = true;
  EditingHabit = new Habit() {
    Id = SelectedHabit!.Id,
    Name = SelectedHabit!.Name,
    Description = SelectedHabit!.Description};
}
```

Add a function that will allow us to save a habit that we have been editing:

```
private void SaveHabit()
{
  GoodHabits.Add(EditingHabit!);
  GoodHabits.Remove(SelectedHabit!);
  IsEditing = false;
  SelectedHabit = null;
}
```

Finally, add a function that will allow us to cancel any edits that we made, should we change our minds:

```
private void CancelEdit()
{
  IsEditing = false;
  EditingHabit = null;
  SelectedHabit = null;
}
```

That should allow us to manipulate the habits in the list! Now, we will add the UI and bind the elements to the functions that we just created!

We'll start by adding the list of habits to the left-hand pane. Copy the following HTML in directly under the <h3>Good Habits</h3> line in the HTML:

```
<ul class="list-group">
    @foreach (var habit in GoodHabits)
    {
        <li class="list-group-item d-flex justify-content-
          between align-items-center">
            <span @onclick="() =>
              ShowDetails(habit)">@habit.Name</span>
            <div>
                <i class="oi oi-eye mr-2 text-primary"
                  @onclick="() => ShowDetails(habit)"></i>
                <i class="oi oi-pencil mr-2 text-primary"
```

```
            @onclick="() => EditHabit(habit)"></i>
          <i class="oi oi-trash text-danger"
            @onclick="() => DeleteHabit(habit)"></i>
        </div>
      </li>
    }
  </ul>
  <button class="btn btn-primary mt-3" @onclick="AddHabit">Add Habit</
  button>
```

You can see that we loop over the list of dummy habits that we included in the code block, binding these habits to the functions that we added to the @code block. We have added buttons to add, edit, and delete the habits in the list.

Finally, we need to add the HTML to display or edit the habit on the right-hand side of the screen. Add the following HTML directly under the <h3>Habit Details</h3> element:

```
@if (SelectedHabit != null)
{
    <div class="card">
        <div class="card-body">
            @if (!IsEditing)
            {
                <h5 class="card-
                  title">@SelectedHabit.Name</h5>
                <p class="card-
                  text">@SelectedHabit.Description</p>
                <button class="btn btn-danger mt-3"
                  @onclick="() => DeleteHabit
                  (SelectedHabit)">Delete Habit</button>
                <button class="btn btn-primary mt-3"
                  @onclick="() => ShowEditForm()">Edit
                  Habit</button>
            }
            else
            {
                <form>
                    <div class="form-group">
                        <label for="edit-habit-
                          name">Name</label>
                        <input type="text" class="form-
                          control" id="edit-habit-name"
                          placeholder="Enter habit name"
                          @bind-value="EditingHabit.Name"
                          />
                    </div>
```

```
<div class="form-group">
    <label for="edit-habit-
      description">Description</label>
    <textarea class="form-control"
      id="edit-habit-description"
      rows="3" @bind=
      "EditingHabit.Description">
      </textarea>
</div>
<button type="submit" class="btn btn-
  primary mt-3" @onclick="() =>
  SaveHabit()">Save</button>
<button type="button" class="btn btn-
  secondary mt-3" @onclick="() =>
  CancelEdit()">Cancel</button>
        </form>
      }
    </div>
  </div>
}
```

There is quite a lot going on here. We have added two forms, one that is shown if the `IsEditing` property is set to `true` and one that is shown if it is not, allowing us to view or edit the habits.

The best way to understand what is happening is to start the project and see what we have created!

Type `dotnet run` into the console and go to the site in a browser. Then, navigate to the GoodHabits page, and you should see the following:

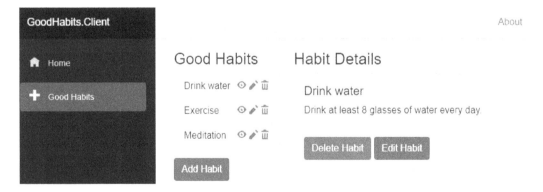

Figure 7.6 – The Good Habits UI

As you can see in the preceding screenshot, we have a list of habits, and clicking on the name of any of these will bring up the "view" form.

The UI should be fairly intuitive, if not very pretty! I think it is a very useful exercise to play with this UI and look at how the various functions and elements we added to the Razor file interact.

So far, we have a functional UI, but there is no connection to that backend yet. Let's work on that now!

Connecting the UI to the backend

Previously, we built the UI and added the functionality to add, delete, view, and update the habits. However, currently, there is no connection to the API that we have built up. Let's remedy that now!

Dealing with CORS issues

CORS stands for **Cross-Origin Resource Sharing**. It's a mechanism that allows web pages to make requests to a different domain than the one that served the web page. CORS is a security feature that helps to prevent unauthorized access to web servers. When a web page attempts to make a cross-origin request, the server it makes the request to must respond with specific headers that allow the request to go through. If the server does not send these headers, the browser will block the request.

We need to configure the API gateway project to accept connections from any origin when in development mode. Please note that this represents a security risk and should not be replicated in production.

Open up `Program.cs` in the `API Gateway` project (*not in the Blazor project – make sure you are in the correct* `Program.cs` *file!!*). Locate the development settings:

```
if (app.Environment.IsDevelopment())
{...}
```

Now, make sure that the following code is included:

```
app.UseCors(policy =>
    policy.AllowAnyOrigin()
        .AllowAnyHeader()
        .AllowAnyMethod()
        );
```

This will ensure that the API will accept requests, but again, this is only suitable in development and not suitable for a production environment!

Logic, including calling the API

We will modify the code that we built previously to connect to the API. If we did our jobs well, and correctly separated the UI elements from the logic, we should be able to make this change by only modifying the code. Let's see how we did!

Start by adding a string constant that points to the API URL:

```
private const string ServiceEndpoint = "http://localhost:5300/
gateway/habits";
```

Make sure to set the port correctly. Also, note that hardcoding a URL is not really good practice, but it will suffice for this demo.

Next, comment out or delete the three dummy habits that we added. From now on, we will get the habits that we stored in our database:

```
private List<Habit> GoodHabits { get; set; } = new List<Habit>();
```

We need to add a hook to fetch the habits from the database. We will tap into the initialization hooks and add the following method:

```
protected override async Task OnInitializedAsync()
{
    httpClient.DefaultRequestHeaders.Add("tenant",
      "CloudSphere");
    GoodHabits = await
      httpClient.GetFromJsonAsync<List<Habit>>($"
      {ServiceEndpoint}");
}
```

This uses `httpClient` to call the API endpoint, which will return the list of habits that are stored in the database.

Before we can add the calls to interact with the add and edit endpoints, we will need to create some classes. Add two private classes as follows:

```
private class CreateHabit
{
    public string Name { get; set; } = default!;
    public string Description { get; set; } = default!;
    public int UserId { get; set; }
}

private class UpdateHabit
{
    public string Name { get; set; } = default!;
    public string Description { get; set; } = default!;
}
```

Next, modify the `AddHabit` method to interact with the API:

```
private async Task AddHabit()
{
    var newHabit = new CreateHabit()
    {
        Name = "New Habit",
        Description = "Enter a description here",
        UserId = 101
    };

    var response = await httpClient
      .PostAsJsonAsync(ServiceEndpoint, newHabit);
    var habit = await response
      .Content.ReadFromJsonAsync<Habit>();
    // Add the new habit to the list of habits
    GoodHabits.Add(habit!);
    SelectedHabit = habit;
}
```

Add a line to the `DeleteHabit` method:

```
private void DeleteHabit(Habit habit)
{
    httpClient.DeleteAsync($"{ServiceEndpoint}/
      {habit.Id}");
    GoodHabits.Remove(habit);
    if (SelectedHabit == habit)
    {
        SelectedHabit = null;
    }
}
```

And finally, modify the `SaveHabit` method to include the required interactions:

```
private void SaveHabit()
{
    httpClient.PutAsJsonAsync($"{ServiceEndpoint}/
      {EditingHabit!.Id}",
            EditingHabit);
    GoodHabits.Add(EditingHabit!);
    GoodHabits.Remove(SelectedHabit!);
    IsEditing = false;
    SelectedHabit = null;
}
```

That's done! With just a few small changes, we are now hooked up to the API and the database. We have now created the outline of a fully functioning SaaS application!

To prove this, go into the **Run and Debug** menu and execute the **Run All** compound task that we created in *Chapter 6*. This will start both the microservice projects, the API gateway, and the client project. Then, navigate to the client in the browser. You should see the contents of your habits database on the screen! We are much closer to completing our SaaS application!

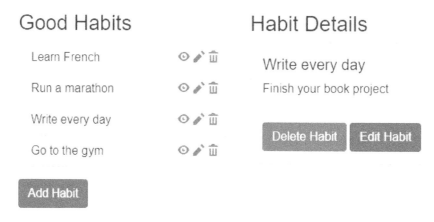

Figure 7.7 – The Good Habits UI, connected to the gateway API

Building usable UIs

The example that we have built here is a good technical demonstration of the essential techniques to incorporate the SaaS backend with a UI, but I think it's safe to say that it lacks a bit of flair. These days, the average consumer expectations of a UI for a SaaS app are very high – essentially, customers will demand a very similar experience to a native application but running in a browser.

To make a UI look more modern and responsive, we would typically use an off-the-shelf responsive UI framework. In addition to expecting the UI to perform like a desktop app, there is generally also an expectation that the UI should work on a tablet and a phone. Finally, not everyone shares the same abilities, and there are standard methods to endure that your UI is accessible to differently abled persons. In this section, we will look into all this, starting with responsive UI frameworks.

Responsive UI frameworks

Responsive UI frameworks are a collection of pre-built components, styles, and scripts designed to help developers create responsive and adaptive web applications with ease. These frameworks provide developers with a set of tools that can be used to build UIs that automatically adjust to different devices and screen sizes, ensuring a consistent and enjoyable UX across various platforms.

The use of responsive UI frameworks can significantly streamline the development process by providing ready-to-use components and a cohesive design system. This enables developers to focus more on an application's functionality and less on the complexities of building responsive layouts from scratch. Additionally, these frameworks typically adhere to well-established design principles and best practices, ensuring that the final product is not only visually appealing but also accessible and user-friendly.

Bootstrap is an open source and widely used CSS framework developed by Twitter. It simplifies the process of creating responsive and mobile-first websites by providing a comprehensive set of reusable components, such as buttons, forms, and navigation elements. Bootstrap also includes a responsive grid system based on the **Flexible Box Layout Module** (**Flexbox**), making it easy to create fluid layouts that adapt to different screen sizes. With its extensive documentation and large community, Bootstrap remains a popular choice among developers.

Foundation, created by ZURB, is another popular responsive frontend framework that focuses on providing a robust and flexible foundation to build custom web applications. It offers various pre-built components, a responsive grid system, and a modular architecture that enables developers to use only the components they need for their projects. Foundation is known for its performance optimizations and compatibility with a wide range of devices, making it a suitable choice for complex projects that require advanced customization and performance.

Material-UI is a popular React-based UI framework that implements Google's Material Design guidelines. Material-UI provides a consistent and modern look and feel for web applications, ensuring that UIs are both visually appealing and easy to navigate. It includes a set of pre-built components, a responsive layout system, and a theming system that allows for easy customization. By following Material Design principles, Material-UI helps developers create UIs that adhere to established usability standards.

While Flexbox is a powerful CSS layout module that simplifies the design of flexible and responsive layouts for web pages, it is not a full-fledged responsive UI framework. Instead, it is a valuable tool that can be used in conjunction with other frameworks to create adaptive layouts. Many responsive UI frameworks, such as Bootstrap and Foundation, incorporate Flexbox as part of their grid systems, leveraging its capabilities to create versatile and fluid layouts for their components.

It is definitely recommended to choose one of these to integrate with your project. It would be an excellent learning exercise to apply one of these frameworks to the demo UI that we created previously!

Responsive design techniques

Simply installing one of the responsive frameworks described previously is not quite enough. You will need to master some of the techniques to get the most out of them.

In this section, we will quickly explore the fundamental responsive design techniques, including fluid grids, flexible images, and media queries. By understanding and applying these techniques in conjunction with responsive UI frameworks, you can build usable UIs that provide a consistent and enjoyable experience across various platforms.

Fluid grids form the backbone of responsive design by enabling flexible layouts that adjust dynamically, based on screen size. Instead of using fixed-width units such as pixels, fluid grids use relative units such as percentages to define the width of elements. This ensures that the layout automatically scales and reflows as the viewport changes. For example, when working with Bootstrap, you can create a fluid grid by utilizing its predefined grid classes, such as `.container`, `.row`, and `.col`. These classes enable you to define a responsive grid structure that adapts to different screen sizes.

Flexible images ensure that media content, such as images and videos, also scales and adapts to different screen sizes. By setting the max-width property of images to 100%, images will automatically scale down to fit the width of their containing element, preventing them from overflowing and disrupting the layout. When working with frameworks such as Foundation or Material-UI, you can use their built-in classes or components to handle image scaling, ensuring that your media content remains responsive across various devices.

Media queries are a powerful CSS feature that allows you to apply different styles based on the characteristics of a user's device, such as screen size, resolution, or orientation. By using media queries, you can define breakpoints at which your layout and styles change, ensuring that your UI remains usable and visually appealing at different screen sizes. Most responsive UI frameworks, such as Bootstrap and Material-UI, provide predefined media queries and breakpoints that you can use or customize to suit your specific needs.

By combining these core responsive design techniques with the features provided by responsive UI frameworks, you can create web applications that are not only visually appealing but also highly functional and adaptable to various devices and screen sizes. This ultimately contributes to building usable UIs that enhance the overall UX and cater to diverse user needs and preferences.

Designing for different screen sizes

These days, it is impossible to predict what device a user will choose to use to access your SaaS application. As a result, it's crucial to consider different screen sizes and resolutions when designing responsive web applications. By doing so, you can ensure that your UIs look and function well across different devices, providing a consistent and enjoyable UX.

When designing for different screen sizes, it's essential to follow a few key guidelines and best practices:

- It can be very helpful to take a mobile-first approach. Start by designing your layout and UI for smaller screens, such as smartphones, and then progressively enhance the design for larger screens. This approach ensures that your application remains functional and visually appealing on smaller devices, while taking advantage of the additional screen real estate on larger ones.

- As mentioned earlier, responsive UI frameworks such as Bootstrap, Foundation, and Material-UI provide pre-built components, grid systems, and predefined media queries that make it easier to create adaptive layouts for various screen sizes. Leveraging these frameworks can significantly streamline the development process and ensure that your UIs remain consistent and functional across different devices. Remember to apply best practices and use good responsive design techniques!

- Always test on multiple devices and screen sizes. During the development process, test your application on a variety of devices and screen sizes to identify potential issues and ensure a consistent UX. You can use device emulators, browser developer tools, or physical devices to test your application's responsiveness and make any necessary adjustments.

- Optimize your application's performance for different devices, as slower load times and inefficient resource usage can significantly impact the UX, especially on mobile devices. Consider factors such as image optimization, code minification, and lazy loading to improve the performance of your application across different screen sizes.

By following these guidelines and best practices, you can create responsive web applications that provide a consistent and enjoyable UX across various devices and screen sizes. Utilizing responsive UI frameworks and responsive design techniques will ensure that your UIs adapt seamlessly, catering to the diverse needs and preferences of your users.

Accessibility

Creating accessible and inclusive web applications is an essential aspect of responsible and empathetic design. By considering the needs of differently abled users, you ensure that your SaaS applications provide equal access and opportunities for everyone, fostering a more inclusive online environment.

Embracing accessibility and inclusivity in your web applications is really important for a few reasons. Doing so gives a wider audience reach. By making your application accessible, you cater to a larger audience, including differently abled users, who might otherwise face barriers when interacting with your application.

Generally speaking, a UI that is designed to be accessible will result in a better overall experience in general. Accessible design principles often result in better usability for all users, as they promote clear and intuitive interfaces that are easy to navigate and understand.

Finally, it is the right thing to do. We should strive to make the internet, and indeed the world, a more inclusive place. If we can make a very tiny difference by taking the time to make our applications accessible, then we should!

To create more accessible web applications, it's important to follow established accessibility standards and guidelines, such as the **Web Content Accessibility Guidelines** (**WCAG**) and the **Americans with Disabilities Act** (**ADA**) Standards for Accessible Design. These guidelines provide a framework to ensure that your application is usable and accessible for differently abled users.

There are several tools and techniques available to address common accessibility challenges and improve the UX for differently abled users:

- **Screen readers**: These assistive technologies convert text and other onscreen content into speech or braille, helping visually impaired users to access and navigate web applications. Ensure that your application's content is structured semantically, with proper use of headings, landmarks, and alternative text for images to support screen reader users.

- **Keyboard navigation**: Some users may rely solely on a keyboard to navigate web applications. Ensure that your application supports keyboard navigation by providing visible focus indicators, logical tab order, and keyboard-accessible interactive elements.

- **Color contrast**: Users with visual impairments or color blindness may have difficulty perceiving content with low contrast. Ensure that your application's color scheme and design elements adhere to the recommended contrast ratios, as specified by the WCAG.

- **Accessible forms**: Users with cognitive or motor impairments may struggle with complex forms and input fields. Simplify forms, provide clear labels, and use proper input types to make it easier for all users to interact with your application.

- **Accessible Rich Internet Applications (ARIA)**: This set of attributes helps enhance the accessibility of dynamic content and advanced UI controls. Use ARIA attributes to provide additional information about the structure and functionality of your application, ensuring that assistive technologies can interpret and interact with it correctly.

By considering the needs of differently abled users and implementing these tools and techniques, you can create web applications that are more accessible, inclusive, and user-friendly. This not only benefits your users but also contributes to the overall success of your application.

Summary

In this chapter, we've covered a lot of ground in terms of designing and building a UI for a SaaS application. We've talked about the importance of UX and how to design for user personas, plan user journeys, and create visually appealing and responsive designs. We've also discussed the importance of UI testing and how to build a simple UI using Blazor.

One of the key takeaways from this chapter is the importance of UX in the development of a SaaS application. A well-designed and intuitive UI can make all the difference in terms of user adoption, retention, and satisfaction. By planning for user personas and user journeys, we can ensure that we're building an interface that meets the needs and expectations of our target audience.

Another important takeaway is the value of using a modern UI framework such as Blazor. By using Blazor, we can take advantage of the power and flexibility of .NET to build rich, interactive, and responsive UIs that can communicate effectively with backend APIs. Blazor allows us to use C# and .NET skills to build web applications that run in a browser, using WebAssembly to execute .NET code on the client side.

We've also covered some key best practices to build a UI, including designing for accessibility, using responsive design, optimizing performance and load times, and providing feedback and testing for users. These are all essential elements of a well-designed and user-friendly UI.

In the second part of the chapter, we delved into how to connect a Blazor UI to a backend API. We discussed how to configure the Blazor client to communicate with the API, how to define data models, and how to retrieve and update data. We also talked about the importance of error handling, testing, and debugging, ensuring that our application is robust and reliable.

By following the steps outlined in this chapter, you will have a good understanding of how to design and build a UI for a SaaS application and how to connect that interface to a backend API. You will be equipped with the tools and knowledge needed to build an intuitive, visually appealing, and responsive UI that meets the needs of your target audience, while also being performant and reliable.

In conclusion, the UI is a critical component of any SaaS application, and designing and building a great UI requires a combination of technical and creative skills. By following the best practices and guidelines covered in this chapter, you will be well on your way to building a UI that is intuitive, engaging, and effective, helping you to achieve your business goals.

In the next chapter, we will talk about authentication and authorization, with specific reference to how this will affect a multi-tenant, microservice SaaS application!

Further reading

- Build beautiful web apps with Blazor: `https://dotnet.microsoft.com/en-us/apps/aspnet/web-apps/blazor`

- VS Code and Blazor WASM: `https://dev.to/sacantrell/vs-code-and-blazor-wasm-debug-with-hot-reload-5317`

- Call a web API from ASP.NET Core Blazor: `https://learn.microsoft.com/en-us/aspnet/core/blazor/call-web-api?view=aspnetcore-7.0&pivots=server`

- Blazor WebAssembly HttpClient: `https://code-maze.com/blazor-webassembly-httpclient/`

Questions

1. What is Blazor, and how does it allow us to create web applications using C# and .NET?

2. How can we automatically generate a client that connects to a .NET web API from a Blazor application?

3. How can we consume data from a .NET web API in a Blazor application, and what are some best practices to handle errors and exceptions?

4. What is a user persona?

5. Why is it important to map a user journey before starting to build a UI?

8

Authentication and Authorization

Authentication and authorization are fundamental concepts in software development and have particular importance in the context of **software-as-a-service** (**SaaS**) applications. In a SaaS context, all of the users' data is typically stored remotely and is only as secure as the authentication and authorization mechanisms. These mechanisms help to ensure that users can interact with the application in a secure and controlled manner and that sensitive data and resources are protected from unauthorized access.

In this chapter, we will explore the key concepts and best practices for implementing authentication and authorization, with a focus on doing so in a SaaS application. Of course, we will focus on the Microsoft stack, but the principles that we cover should be applicable to most modern options for web development. We will begin by discussing the differences between authentication and authorization and how these mechanisms work together to provide a secure environment for users and their data.

We will then move on to explore some of the technical considerations for implementing authentication and authorization in a SaaS application and will consider some of the specific challenges that are faced by developers of such applications. In particular, we'll consider how multi-tenancy (as discussed in *Chapter 3*) and microservice architecture (as discussed in *Chapter 6*) impact the security landscape.

Throughout an application's life cycle, users will come and go, and at times they will change their roles within the application. We'll take a look at how to manage the changing and hopefully growing user base.

Finally, we'll work on a practical example, where we will add authentication and authorization to our demo application, using the skills we have covered in this chapter to build a robust model for security that could be scaled up for use in a real-world application.

By the end of this chapter, you will have a clear understanding of the fundamental concepts and best practices for implementing authentication and authorization in a SaaS application. You will also have a deeper appreciation for the importance of getting these mechanisms right and how doing so can help to protect valuable data and resources, as well as build trust and confidence among users!

The main topics covered in this chapter are the following:

- An overview of authentication and authorization

- Challenges arising from core SaaS concepts, such as multi-tenancy and microservices

- How to manage users, roles, and permissions

- Adding authentication and authorization to our demo app

Technical requirements

All code from this chapter can be found at `https://github.com/PacktPublishing/Building-Modern-SaaS-Applications-with-C-and-.NET/tree/main/Chapter-8`.

What are authentication and authorization

Before diving into the implementation details, let's take a moment to understand the fundamental concepts of authentication and authorization using a real-world analogy. Imagine that your application is like a secured building, and the various resources or actions within the application are represented by rooms inside the building. To ensure the security of the building and its contents, access to the rooms is controlled by a two-step process: authentication and authorization.

Authentication is the process of verifying the identity of a person or entity attempting to access the building, much like presenting an ID card to the security guard at the entrance. In the context of an application, authentication involves confirming that a user is who they claim to be, typically through the use of a username and password. This is the first step in ensuring the security of your application.

Once a user's identity has been authenticated and they have been allowed to enter the building, the next step is to determine what they are allowed to do within the building. This is where authorization comes into play. Authorization is the process of granting or denying access to specific resources or actions based on the authenticated user's permissions, just like an access card or key you receive after your identity is verified. These permissions are usually assigned through roles or claims and can be as simple or complex as your application requires.

This idea of a secured building with ID required to get into the building, and then once in, keycard access for certain parts of the building, is a very useful analogy for authentication and authorization that you should keep in mind as we dive a little deeper into these concepts!

Authentication

We'll start by delving deeper into the various aspects of authentication, exploring different forms and methods, their implementation in .NET, and best practices. By understanding the nuances of authentication and how to properly implement it, you can build a robust foundation for protecting your application and its users as the rest of the application takes shape.

While most often, we think of a username and password for authentication, there are a few ways to approach this. Alternatives to username and password include token-based authentication, **multi-factor authentication (MFA)**, and **single sign-on (SSO)**. We will take a look at how to implement these authentication methods, with a focus on how this works in .NET-based applications.

We will also cover some other important topics, such as securely storing passwords and secrets, as well as best practices for enforcing strong password policies and implementing account lockout policies.

Forms of authentication

In the world of application security, there are several forms of authentication to verify the identity of someone wishing to use the application. Each method has its own advantages and limitations.

The most common form of authentication is a simple username and password system, with which we are all familiar! This method relies on users keeping their passwords confidential and choosing strong, complex passwords to reduce the risk of unauthorized access, which can be a fairly significant flaw in a security system!

Using MFA can help to mitigate this. MFA requires users to provide two or more forms of authentication to verify their identity, which can greatly improve a system's overall security.

In an enterprise setting, it is common for the organization to use SSO. This allows users to access multiple related applications or services using a single set of credentials. The big advantage of this is that the organization has a lot of control over the security setup. For example, they are able to insist on passwords of a certain complexity or enforce MFA.

Implementing authentication in .NET

In this subsection, we will explore how to implement various authentication methods in .NET using ASP.NET Core Identity and integrating with external authentication providers. We will discuss the configuration and customization of these methods to align with your application's requirements.

Later in this chapter, we will use these techniques to add authentication to our demo application.

ASP.NET Core Identity

ASP.NET Core Identity is a flexible and extensible framework that provides a secure way to manage user authentication and authorization. It includes features such as password hashing, two-factor authentication, and support for external authentication providers. To get started with ASP.NET Core Identity, you'll need to install the necessary NuGet packages and configure your application by following these steps:

1. Install the required NuGet packages with the following command:

    ```
    dotnet add package Microsoft.AspNetCore.Identity.
    EntityFrameworkCore
    dotnet add package Microsoft.AspNetCore.Identity.UI
    ```

2. Update your application's `DbContext` to inherit from `IdentityDbContext`, which includes the necessary `Identity` tables for storing user information.

3. Register the `Identity` services in the `ConfigureServices` method of your `Startup` class by adding `services.AddIdentity` and `services.AddAuthentication`.

4. Configure middleware for `Identity` and authentication by adding `app.UseAuthentication` and `app.UseAuthorization` in the `Configure` method of your `Startup` class.

5. Modify your views and controllers to include the necessary authentication functionality, such as login, registration, and logout actions.

You will see the preceding steps in action when we add authentication to the demo application later in this chapter.

Integrating with external authentication providers

To enhance the user experience and security of your application, you may want to integrate with external authentication providers, such as OAuth 2.0 and OpenID Connect, and social logins, such as Microsoft, Google, Facebook, or Twitter.

OAuth 2.0 is an authorization framework that enables your application to obtain limited access to user accounts on an external service, while **OpenID Connect (OIDC)** is an authentication layer built on top of OAuth 2.0 that provides a secure way to authenticate users and obtain their basic profile information.

To implement OAuth 2.0 and OIDC in your .NET application, you can use the `Microsoft. AspNetCore.Authentication.OpenIdConnect` package. This package includes middleware to handle the OIDC authentication flow, such as obtaining an authorization code, exchanging it for an access token, and validating the token.

Doing so is outside the scope of the demo application, but it may be a useful exercise to try and add this yourself!

ASP.NET Core Identity also supports integration with popular social login providers such as Google, Facebook, and Twitter. To implement social logins in your application, follow these steps:

1. Register your application with the desired social login provider to obtain a client ID and client secret.

2. Install the corresponding NuGet package for the social login provider, such as `Microsoft. AspNetCore.Authentication.Google`, `Microsoft.AspNetCore. Authentication.Facebook`, or `Microsoft.AspNetCore.Authentication. Twitter`.

3. Configure the social login provider in the `ConfigureServices` method of your `Startup` class by adding `services.AddAuthentication().Add[ProviderName]` and pass the client ID and client secret obtained earlier.

4. Update your login view to include buttons or links for each social login provider.

Every user of your application will be different and will have different preferences for logging into your application. By implementing various authentication methods in .NET and integrating with external providers, you can create a secure and user-friendly authentication experience for your SaaS application.

Securely storing passwords and secrets

Protecting sensitive information such as user passwords and application secrets is crucial to maintaining the security and integrity of your SaaS application. In this section, we will discuss techniques for securely storing passwords and secrets in your .NET application.

Password hashing and salting

When storing a user's password in a database, or anywhere for that matter, storing that password in 'plaintext' is always a huge mistake that will compromise your application security. Instead, passwords should be hashed and salted before being stored in the database.

Plaintext refers to storing the password as the user enters it. So if the password is 'Passw0rd1', then that string is the plaintext representation of that password. Hashing is a one-way function that transforms the password into a fixed-length string of characters, while salting involves adding a random value (which is known as 'the salt') to the password before hashing to prevent attacks using precomputed tables.

ASP.NET Core Identity automatically handles password hashing and salting using the **Password-Based Key Derivation Function (2PBKDF2)** algorithm. You can configure the hashing settings, such as the number of iterations, through the `IdentityOptions.Password` settings in the `ConfigureServices` method of your `Startup` class.

Leveraging the built-in identity tools in .NET offers significant advantages. Developing custom identity providers can be both challenging and error-prone. Utilizing well-established and battle-tested solutions is always the preferred approach!

Securely managing API keys and other secrets

In addition to users' passwords, your application may also rely on sensitive information such as API keys, connection strings, or encryption keys. Again, storing these secrets in plaintext or hardcoding them in your source code is a mistake that can expose your application to security risks and should be avoided at all costs!

In much the same way that the built-in .NET Core Identity services should be used, pre-existing and battle-tested tools and technologies should be used to manage the application secrets. Here are some best practices that should be your go-to approach!

- **.NET Secret Manager:** The Secret Manager tool allows you to store sensitive information during the development process without including it in your source code. Secrets are stored in a separate JSON file on your development machine and can be accessed using the `IConfiguration` interface in your application. This is a great way to keep development secrets separate from the secrets of the production environment.

- **Environment variables**: Storing secrets in environment variables helps to keep them separate from your application code and allows for easy configuration changes. In production environments, consider using a centralized configuration management solution to manage environment variables and secrets securely.

- **Azure Key Vault**: Azure Key Vault is a cloud-based service for securely storing and managing secrets, keys, and certificates in a production environment. By integrating your .NET application with Azure Key Vault, you can centralize the management of sensitive information and enforce strict access controls. To access secrets stored in Azure Key Vault, use the `Microsoft.Extensions.Configuration.AzureKeyVault` package and configure it in your `Startup` class.

By securely storing passwords and application secrets, you help protect your application and its data from unauthorized access and potential security breaches. Adopting these best practices will ensure that sensitive information remains confidential and secure in your .NET-based SaaS application.

Authentication best practices

Implementing a secure and effective authentication process is crucial for the overall security of your SaaS application. By following best practices, you can enhance the user experience, improve security, and minimize the risk of unauthorized access.

Enforcing strong password policies

To protect against weak or easily guessed passwords, enforce strong password policies in your application. ASP.NET Core Identity allows you to configure password requirements, such as minimum length, complexity, and character types. Consider the following guidelines for strong password policies:

- Require a minimum length of at least 12 characters; more is better. Passwords that are too short can easily fall to brute force attacks.

- Enforce the use of a mix of character types, including uppercase and lowercase letters, numbers, and special characters. Increasing the number of characters to choose from makes the passwords harder to guess.

- Disallow easily guessable passwords or common patterns, such as "password123" or "qwerty."

- Do not require regular password changes. It used to be considered good practice to require users to frequently change their passwords, but this is no longer the case as frequent changes can lead to weaker passwords, as users struggle to remember their ever-changing passwords.

- Encourage MFA. MFA adds an extra layer of security by requiring additional verification methods beyond the password, such as a one-time code, a hardware token, or biometric data.

Monitoring and auditing authentication events

Monitoring and auditing authentication events can help you identify suspicious activity, detect unauthorized access attempts, and maintain a secure environment for your SaaS application. ASP. NET Core Identity provides built-in support for logging authentication events, which should always be used over writing your own implementation.

Consider implementing the following monitoring and auditing practices:

- Log all authentication events, including successful logins, failed login attempts, password changes, and account lockouts.

- Regularly review authentication logs to identify unusual patterns, such as multiple failed login attempts from the same IP address or unusual login times. This process could be automated.

- Implement real-time monitoring and alerting for critical authentication events, such as repeated failed login attempts or unauthorized access to sensitive resources.

- Ensure that logs are stored securely and retained for a sufficient period to support incident response and forensic analysis.

Implementing account lockout policies

Account lockout policies can help protect against brute-force attacks, where an attacker repeatedly attempts to guess a user's password. ASP.NET Core Identity supports account lockout functionality, allowing you to lock out a user's account after a specified number of failed login attempts.

Consider the following guidelines when implementing account lockout policies:

- Set a reasonable threshold for the number of failed login attempts before locking an account, such as 3–5 attempts.

- Determine an appropriate lockout duration, balancing security concerns with user experience. This could range from a few minutes to several hours, depending on your application's requirements.

- Implement a mechanism for users to unlock their accounts, such as by contacting support, resetting their password, or using a secondary authentication factor.

- Monitor account lockout events to identify potential brute-force attacks or other security threats.

It is often the case with the development that the team can, to a certain extent, pick and choose which best practices they want to strictly adhere to. This is fine for the most part but is decidedly not so when it comes to authentication. Commonly understood best practices should always be followed, and out-of-the-box implementations are always preferred over in-house tools. By keeping the preceding best practices in mind from the start of the development process, we can be sure that our SaaS application is as secure as possible!

Authorization

We have covered authentication in detail; now, it is time to move on to authorization. Authorization involves determining what actions and resources an authenticated user is allowed to access within your application.

We will begin by discussing the core concepts of authorization, such as **role-based access control (RBAC)**, **claims-based access control (CBAC)**, and **attribute-based access control (ABAC)**. Next, we will look into the implementation of authorization in .NET using ASP.NET Core Authorization Policies, Role, and Claims Management with ASP.NET Core Identity and Custom Authorization Middleware and Filters.

Finally, we will discuss best practices for authorization, including the **principle of least privilege (POLP)**, **separation of duties (SoD)**, and regular auditing and monitoring of access controls.

Understanding authorization concepts

Let's start by looking at the core concepts of authorization, which involve determining the actions and resources a user is allowed to access within an application. By understanding these concepts, you can create a secure and efficient access control system for your SaaS application.

Role-Based Access Control (RBAC)

RBAC is an approach to authorization that assigns users to specific roles, which in turn grants them permission to perform certain actions or access specific resources. RBAC simplifies the management of access control by allowing you to define and manage permissions at the role level rather than assigning permissions directly to individual users.

Examples of roles in a SaaS application might include "administrator," "manager," and "user," each with different levels of access to application resources and functionality.

RBAC is typically used when managing permissions for groups of users with similar responsibilities, making it easier to grant and revoke access to resources based on predefined roles.

Claims-Based Access Control (CBAC)

CBAC is an alternative approach to authorization that focuses on claims, which are pieces of information about a user, such as their name, role, or other attributes. In CBAC, permissions are granted based on the user's claims rather than their role.

This approach allows for more fine-grained access control and can provide a more flexible and dynamic authorization system compared to RBAC. Claims can be issued by your application or external authentication providers, such as social logins or enterprise identity systems such as **Azure Active Directory (Azure AD)**.

Claims-based access control is preferred when you need more fine-grained and dynamic control over user access.

Attribute-Based Access Control (ABAC)

ABAC is a more advanced approach to authorization that evaluates a set of attributes associated with a user, resource, action, and environment to determine whether access should be granted. ABAC enables context-aware access control decisions based on a rich set of attributes and can support complex access control policies.

In an ABAC system, rules or policies are defined using a policy language, such as **eXtensible Access Control Markup Language (XACML)**. These rules are then evaluated by a **policy decision point (PDP)** to determine whether access should be granted or denied.

ABAC is preferred when you require a highly granular and context-aware authorization system that considers multiple attributes, such as user characteristics, resource attributes, and environmental factors, to make access control decisions.

Implementing authorization in .NET

Later, we will build authorization into our demo app. First, we will discuss how to implement various authorization concepts in .NET using ASP.NET Core Authorization Policies, Role and Claims Management with ASP.NET Core Identity, and Custom Authorization Middleware and Filters.

ASP.NET Core Authorization Policies

ASP.NET Core provides a powerful and flexible authorization framework that allows you to define and enforce access control policies based on roles, claims, or custom logic. To implement authorization policies in your .NET application, follow these steps:

1. Define authorization policies in the `ConfigureServices` method of your `Startup` class by adding `services.AddAuthorization` and configuring policy options using the `AddPolicy` method. You can specify requirements based on roles, claims, or custom rules.

2. Apply authorization policies to your controllers or action methods using the `[Authorize]` attribute with the specified policy name. This attribute ensures that only users who meet the policy requirements can access the protected resources.

3. If needed, create custom authorization handlers and requirements to implement complex authorization logic or integrate with external systems. Register your custom handlers in the `ConfigureServices` method of your Startup class.

Role and claims management with ASP.NET Core Identity

ASP.NET Core Identity provides built-in support for managing roles and claims, making it easy to implement RBAC and CBAC in your application. To use roles and claims with ASP.NET Core Identity, follow these steps:

1. Enable role management in your application by updating `DbContext` to inherit from `IdentityDbContext` with a role type such as `IdentityRole`.

2. Add role management services to the `ConfigureServices` method of your `Startup` class by calling `services.AddIdentity` with the `AddRoles` method.

3. Use the `RoleManager` and `UserManager` classes in your application to create, update, and delete roles, assign roles to users, and manage claims associated with users.

4. Protect your application resources using the `[Authorize]` attribute with roles or policy-based authorization, as discussed in the previous section.

Custom authorization middleware and filters

In some cases, you may need to implement custom authorization logic that goes beyond roles, claims, and policies. ASP.NET Core allows you to create custom middleware and filters to perform additional authorization checks or enforce access control at a global level.

To create custom middleware, define a new class that implements the `IMiddleware` interface and perform your authorization checks in the `InvokeAsync` method. Register your custom middleware in the `Configure` method of your `Startup` class by calling `app.UseMiddleware`.

To create a custom authorization filter, define a new class that implements the `IAuthorizationFilter` or `IAsyncAuthorizationFilter` interface, and perform your authorization checks in the `OnAuthorization` or `OnAuthorizationAsync` method. Register your custom filter globally in the `ConfigureServices` method of your `Startup` class by adding it to the `services.AddControllers` or `services.AddMvc` options.

Integrating with external authorization services

In some scenarios, you may wish to integrate your .NET application with external authorization services, such as Azure AD, Azure AD B2C, or OAuth 2.0 resource servers, to manage access control for your users. In this subsection, we will discuss how to integrate your application with these services.

Azure AD and Azure AD B2C

Azure AD is a cloud-based **identity and access management (IAM)** service provided by Microsoft. Azure AD allows you to centralize the management of users, groups, and access control for your application. Azure AD B2C is a related service that provides consumer-focused identity management, allowing you to implement SSO and MFA for your application users.

To integrate your .NET application with Azure AD or Azure AD B2C, follow these steps:

1. Register your application in the Azure portal, and configure your application to use Azure AD or Azure AD B2C for authentication and authorization.

2. In your .NET application, add the `Microsoft.Identity.Web` package and configure the authentication services in the `ConfigureServices` method of your `Startup` class by calling `services.AddAuthentication` and `services.AddMicrosoftIdentityWebApp`.

3. Protect your application resources using the `[Authorize]` attribute with roles, policies, or custom authorization logic, as discussed in the previous sections.

OAuth 2.0 scopes and resource servers

OAuth 2.0 is an industry-standard protocol for authorization, which allows you to grant third-party applications access to your resources on behalf of a user without sharing their credentials. In the context of OAuth 2.0, your .NET application may act as a resource server, which hosts protected resources and requires valid access tokens for authorization.

To integrate your .NET application with an OAuth 2.0 authorization server, follow these steps:

1. Register your application with the authorization server, and configure it to use OAuth 2.0 for authentication and authorization.

2. In your .NET application, add the appropriate OAuth 2.0 or OpenID Connect middleware package, such as `Microsoft.AspNetCore.Authentication.OAuth` or `Microsoft.AspNetCore.Authentication.OpenIdConnect`, and configure the authentication services in the `ConfigureServices` method of your `Startup` class.

3. Define and enforce access control policies based on OAuth 2.0 scopes or claims by implementing custom authorization logic, as discussed in the previous sections.

By integrating your .NET application with external authorization services, you can take advantage of centralized IAM, SSO, MFA, and other advanced security features to protect your application resources and provide a seamless user experience.

Authorization best practices

To ensure a secure and efficient access control system for your SaaS application, it is essential to follow authorization best practices. In this subsection, we will discuss some of the most important best practices to keep in mind while implementing authorization in your application.

Principle of least privilege

The POLP is a fundamental security concept that dictates that users should be granted the minimum level of access necessary to perform their tasks. By adhering to this principle, you can minimize the potential damage caused by unauthorized access or compromised user accounts. To implement POLP, ensure that you do the following:

- Assign users to the least privileged roles or create custom roles with the minimum required permissions.

- Regularly review and update user permissions to ensure that they align with their current responsibilities.

- Implement fine-grained access controls using claims or attributes when necessary to further restrict access to sensitive resources.

Separation of duties

SoD is another important security concept that involves dividing critical tasks and responsibilities among multiple users or roles to prevent any single user from having excessive access or control. To implement SoD in your application, ensure that you perform the following:

- Define distinct roles for different tasks and responsibilities, and assign users to these roles based on their job functions.

- Implement checks and balances, such as requiring multiple approvals for critical actions or using different roles for data entry and data validation.

- Regularly audit and monitor user activity to ensure that SoD is maintained and identify any potential violations or conflicts.

Regular auditing and monitoring of access controls

Continuously monitoring and auditing your access control system can help you identify potential security risks, ensure that user permissions are up-to-date, and detect unauthorized access or abuse. To implement regular auditing and monitoring of access controls, consider the following practices:

- Maintain detailed logs of all authorization events, such as role or permission changes, access attempts, and policy evaluations.

- Regularly review these logs to identify unusual patterns or potential security risks, such as users with excessive permissions, unauthorized access attempts, or policy violations.

- Implement real-time monitoring and alerting for critical authorization events or anomalies and promptly investigate and address any identified issues.

By following these authorization best practices, you can create a secure and efficient access control system for your SaaS application, protecting your valuable resources and data while ensuring a seamless user experience.

The synergy of authentication and authorization

In a robust and secure SaaS application, authentication and authorization work hand in hand to protect your resources and data. While authentication verifies the identity of a user, confirming they are who they claim to be, authorization determines what actions and resources the authenticated user is allowed to access. By effectively combining these two concepts, you can create a strong and comprehensive access control system for your application.

Integrating authentication and authorization effectively in your .NET application involves using technologies such as ASP.NET Core Identity, OAuth 2.0, and Azure AD. These technologies provide a seamless experience for your users while ensuring proper access control. By following best practices for both authentication and authorization, you can minimize potential security risks and maintain the integrity of your SaaS application.

A well-implemented access control system that synergistically combines authentication and authorization not only provides a secure environment for your SaaS application but also helps create a seamless and efficient user experience, contributing to the overall success of your application.

Criticality of secure authentication and authorization

In the rapidly evolving world of SaaS application development, it's crucial to prioritize security and privacy from the very beginning. By investing time and resources in getting authentication and authorization right upfront, you're not only protecting your application and users but also building a strong foundation for future growth and adaptability.

One of the key reasons to emphasize proper authentication and authorization is the potential impact of security breaches. Data leaks, unauthorized access, and cyberattacks can have severe consequences for both businesses and end-users. The financial costs associated with data breaches, damage to brand reputation, and loss of customer trust can be devastating. By implementing robust security measures, you can significantly reduce the risk of such incidents and the associated liabilities.

Another important factor is compliance with data protection and privacy regulations, such as the **General Data Protection Regulation (GDPR)** in Europe and the **California Consumer Privacy Act (CCPA)** in the USA. These regulations require businesses to implement appropriate security measures to protect user data and privacy. Neglecting to do so can lead to hefty fines and legal ramifications. Proper authentication and authorization mechanisms are essential to demonstrate your commitment to data protection and to comply with these regulations.

However, implementing effective authentication and authorization can be challenging, particularly in complex SaaS environments that may involve multi-tenancy, microservices, and distributed systems. As you scale your application, you'll need to ensure that security measures continue to provide a high level of protection and adapt to changing requirements.

Some of the challenges include managing user identities and access control across various services, securely storing and transmitting sensitive data, and maintaining isolation between tenants. Additionally, you'll need to stay up-to-date with the latest security best practices and emerging threats to ensure your application remains secure in the face of new vulnerabilities and attack vectors.

Investing in robust authentication and authorization upfront is essential for the security, privacy, and success of your SaaS application. By doing so, you'll protect your users, comply with regulations, and build a solid foundation for future growth. While it can be a challenging endeavor, taking the time to get it right will pay dividends in the long run, ensuring the ongoing safety and trust of your customers!

Authentication and authorization are large and complex topics, even on a simple application. With SaaS, we are playing on hard mode, though – we also need to think about how to secure a multi-tenant and microservice application. We will consider these nuances in the next section.

Multi-tenancy and microservices

In this section, we will explore the unique challenges and considerations associated with implementing authentication and authorization in multi-tenant and microservices-based SaaS applications. Multi-tenancy requires special attention to ensure proper tenant identification and isolation, as well as managing tenant-specific roles and permissions. On the other hand, microservices architecture presents its own set of challenges, such as centralized authentication, API gateway access control, and secure service-to-service communication.

Multi-tenancy considerations

There are a few specific considerations that we have to think about as developers of multi-tenant applications.

Tenant identification and isolation

In a multi-tenant SaaS application, correctly identifying and isolating tenants is a critical aspect of ensuring data security and privacy. Tenant identification is the process of determining which tenant a user belongs to when they interact with your application. Tenant isolation ensures that data and resources are securely separated between tenants, preventing unauthorized access or data leaks.

As you will remember from *Chapter 3*, there are several approaches to tenant identification, including using subdomains, URL paths, or custom headers. The chosen method should be consistent and easy to manage. Whichever approach you select, it's important to validate tenant identifiers in every request to ensure that users can only access data and resources belonging to their tenant.

Tenant isolation can be achieved at different levels, such as the database, application, or infrastructure level. For example, you could use separate databases or schemas for each tenant, ensuring data is physically separated. Alternatively, you could use a shared database with row-level security, which enforces tenant isolation at the data access layer. At the application level, you can implement tenant-aware middleware or filters that enforce tenant isolation in every request.

When designing your multi-tenant SaaS application, consider the trade-offs between these approaches in terms of complexity, scalability, and maintainability. By effectively implementing tenant identification and isolation, you can build a secure and compliant SaaS application that safeguards tenant data and resources.

Tenant-specific roles and permissions

SaaS applications are frequently multi-tenant, so it is essential to understand how to manage tenant-specific roles and permissions to ensure that users have the appropriate level of access to resources and functionality within their tenant. This not only helps maintain data security but also provides a tailored user experience for each tenant, as different tenants may require different sets of roles and permissions.

One approach to managing tenant-specific roles and permissions is to extend the existing role and permission model by including a tenant identifier. This way, when you assign roles and permissions to users, you can associate them with a specific tenant. This ensures that users can only perform actions and access resources within their tenant's context.

When implementing tenant-specific roles and permissions, consider the following best practices:

- Define a clear and flexible role hierarchy that can accommodate the needs of different tenants. This may include common roles shared across all tenants, as well as custom roles specific to certain tenants.

- Assign roles and permissions based on the POLP, ensuring users have only the necessary access to perform their tasks.

- Implement a user-friendly interface for tenant administrators to manage roles and permissions within their tenants. This allows tenants to have more control over their users' access levels and simplifies the administration process.

- Regularly review and update tenant-specific roles and permissions to ensure they accurately reflect each tenant's requirements and the application's functionality.

Tenant onboarding and offboarding

As well as users coming and going, tenants will also join the application and (sadly) leave. Tenant onboarding and offboarding are important processes in managing multi-tenant SaaS applications. Properly managing these processes helps ensure a smooth and efficient experience for your tenants while maintaining security and compliance.

Onboarding a new tenant starts with tenant registration, where you gather essential information about the new tenant, such as their organization name, contact information, and any custom configuration options they might require. Next, set up the necessary resources for the tenant, such as databases, schemas, or namespaces, and apply any tenant-specific configurations. After the initial setup, create user accounts, including tenant administrators, and assign appropriate roles and permissions.

In addition to the basic setup, consider applying any tenant-specific branding, integrations, or customizations required for the tenant's unique needs. Finally, provide documentation, tutorials, or other support materials to help the tenant's users get started with your application.

When offboarding a tenant, it's essential to follow a well-defined process to ensure clean and secure decommissioning. Start by providing the tenant with the ability to export their data in a standard format, ensuring they retain access to their information. Once the tenant's data has been exported, proceed to resource cleanup by removing the tenant's resources, such as databases, schemas, or namespaces, and any associated data.

Also, deactivate user accounts and revoke any access tokens or API keys related to the tenant. This step helps prevent unauthorized access to your application or system after the tenant has been offboarded. Finally, document the offboarding process and maintain a record of the decommissioned tenant for auditing and compliance purposes.

Tenant onboarding and offboarding focus on managing the entire life cycle of a tenant within the SaaS application, including creating and removing tenant-specific resources, configurations, and customizations, while user provisioning and de-provisioning primarily deal with individual user account management, such as creating, updating, and deleting user accounts within the context of an existing tenant.

This should give you a good start in addressing the specific challenges involved in securing a multi-tenant application. In the next section, we'll consider the other added complexity that we have introduced – microservices.

Microservices architecture considerations

Microservice architectures offer significant advantages, as we have discussed in an earlier chapter, but they do come at the cost of some additional complexity. In this section, we will discuss some of the additional complexities that come from working with microservices.

Centralized authentication and authorization

In a microservices-based SaaS application, implementing centralized authentication and authorization is important for managing access control across multiple services consistently and efficiently. Centralizing these processes ensures that each service adheres to a uniform security policy and reduces the risk of misconfigurations or inconsistencies that could lead to vulnerabilities.

One common approach to centralizing authentication and authorization in a microservices architecture is by using an IAM service, such as Azure AD or Identity Server. The IAM service acts as a single source of truth for user authentication, and it provides a unified way to manage roles, permissions, and access tokens across all services.

When a user attempts to access a protected resource, the request is first sent to the IAM service for authentication. If the user is successfully authenticated, the IAM service generates an access token, which typically includes the user's identity, roles, and permissions. This access token is then passed along with subsequent requests to other services, allowing each service to authorize the user based on the token's contents.

To implement centralized authentication and authorization in your microservices architecture, consider the following best practices:

- Use an established IAM service or framework that supports industry-standard protocols, such as OAuth 2.0 and OpenID Connect, to facilitate interoperability and simplify integration.

- Secure communication between services and the IAM service using encryption, such as HTTPS or **mutual Transport Layer Security (mTLS)**.

- Implement token validation and caching mechanisms in each service to minimize performance overhead and protect against token tampering or replay attacks.

- Regularly review and update the roles and permissions defined in the IAM service to ensure they accurately reflect the functionality and access requirements of each service.

Please note that implementing such a system is outside the scope of the demo app but would serve as a good project for any readers wishing to continue to advance their understanding of this topic.

API gateway and access control

In a microservices-based SaaS application, an API gateway plays a crucial role in managing and securing access to your services. The API gateway acts as a single entry point for all client requests, providing a unified layer of access control and simplifying the process of managing security across your microservices.

By consolidating access control at the API gateway, you can enforce consistent authentication and authorization policies across all services without duplicating the logic in each individual service. This reduces the complexity of your services, as they can focus on implementing their specific functionality rather than handling access control directly.

When a client sends a request to access a protected resource, the API gateway intercepts the request and performs the necessary authentication and authorization checks. This may involve validating access tokens, verifying user roles and permissions, and applying rate limiting or other security measures. If the request meets the required criteria, the API gateway forwards the request to the appropriate service. Otherwise, the request is denied, and an error message is returned to the client.

To implement access control at the API gateway effectively, consider the following best practices:

- Choose an API gateway solution that supports your authentication and authorization requirements, such as Ocelot, Kong, or Azure API Management. Ensure the solution is compatible with your chosen IAM service and can efficiently handle token validation and permission checks.

- Configure the API gateway to enforce access control policies consistently across all services, including validating access tokens, checking user roles and permissions, and applying rate limiting or other security measures.

- Secure communication between the API gateway and your services using encryption, such as HTTPS or mTLS, to protect against data breaches and man-in-the-middle attacks.

- Monitor and log access attempts at the API gateway level to gain insight into potential security threats and help with auditing and compliance.

By implementing access control at the API gateway, you can enhance the security of your microservices-based SaaS application while simplifying management and ensuring a consistent access control policy across all services.

Service-to-service communication and authentication

In a microservices-based SaaS application, secure communication between services is essential to maintain the confidentiality, integrity, and availability of your system. Service-to-service authentication helps ensure that only authorized services can communicate with each other, protecting your application from unauthorized access or potential security threats.

To implement secure service-to-service communication and authentication, you can leverage a variety of techniques and protocols depending on your application's requirements and architecture. Some common approaches include the following:

- **mTLS**: With mTLS, both the client and server services present TLS certificates during the TLS handshake process, allowing each service to verify the identity of the other. This approach provides strong authentication, encryption, and data integrity, making it a popular choice for securing service-to-service communication in microservices architectures.

- **Token-based authentication**: In this approach, services use access tokens, such as **JSON Web Tokens (JWTs)**, to authenticate themselves when communicating with other services. The access token typically contains information about the service's identity and may include additional claims, such as roles or permissions. To validate the token, the receiving service verifies the token's signature and checks the claims against its access control policies.

- **API keys**: API keys are unique identifiers that can be used to authenticate services when they make requests to other services. API keys are usually pre-shared secrets, meaning they must be securely distributed to each service and kept secret to prevent unauthorized access. To authenticate a request, the receiving service checks the provided API key against a list of valid keys and, if matched, grants access.

When implementing service-to-service communication and authentication, consider the following best practices:

- Choose an authentication method that meets your security requirements and is compatible with your existing infrastructure and services.

- Encrypt communication between services using transport-level security, such as HTTPS or mTLS, to protect data in transit.

- Implement token or API key validation and caching mechanisms to minimize performance overhead and protect against token tampering or replay attacks.

- Regularly rotate and revoke tokens, certificates, or API keys to limit their potential exposure and reduce the risk of unauthorized access.

By implementing secure service-to-service communication and authentication, you can protect your microservices-based SaaS application from unauthorized access and potential security threats, ensuring the confidentiality, integrity, and availability of your system.

Managing users, roles, and permissions

In SaaS applications, managing user access efficiently and securely is extremely important. User provisioning and deprovisioning are essential processes for controlling access to resources and ensuring that only authorized users have the necessary permissions. Let's explore these processes in detail!

User provisioning and deprovisioning

User provisioning is the process of creating, updating, and managing user accounts and their access rights within a system or application. This process typically involves creating user accounts with unique identifiers, such as usernames or email addresses. Once the accounts are created, roles or permissions are assigned to users based on their responsibilities within the organization. Furthermore, enforcing password policies, such as minimum length, complexity, and expiration periods, ensures that user accounts remain secure.

Automated provisioning can be particularly beneficial for larger organizations or when integrating with external identity providers (e.g., Azure AD or OAuth2). By automating the provisioning process, you can reduce manual errors in user account creation and role assignment, improve the onboarding experience for new users, streamline the management of user access across multiple services or applications, and enhance security by ensuring that only authorized users have access to specific resources.

User deprovisioning is the process of revoking a user's access rights when they are no longer required, such as when an employee leaves the company or changes roles. This process typically involves disabling or deleting the user account and revoking any assigned roles or permissions. In some cases, it may also be necessary to archive or transfer any associated data. Logging the deprovisioning process is crucial for auditing and compliance purposes.

Timely and accurate deprovisioning is essential for maintaining security and minimizing the risk of unauthorized access. By implementing a systematic deprovisioning process, you can prevent former employees or contractors from accessing sensitive data or resources, reduce the potential for security breaches caused by orphaned or inactive accounts, streamline the management of user access, and ensure that only current employees have the appropriate permissions. Additionally, a thorough deprovisioning process helps you comply with data protection and privacy regulations that require the prompt removal of user access.

A considered approach to user provisioning and deprovisioning ensures that security is maintained, access rights are accurately managed, and compliance with data protection and privacy regulations is upheld throughout the entire life cycle of a user's interaction with the SaaS application.

This is an important topic, and the processes should be agreed upon and implemented early in the life cycle of the application. By implementing robust processes for both tasks, you can enhance security, maintain compliance, and streamline access management across your application's ecosystem.

Role management and assignment

In SaaS applications, managing roles and assigning them to users is a crucial aspect of access control. Roles define a set of permissions that determine the actions a user can perform within the application. By effectively managing roles and assigning them to users, you can achieve a higher level of security and maintain a clear separation of responsibilities.

Role management involves creating and maintaining a set of roles that represent different levels of access or responsibilities within your application. These roles should be designed to reflect the various tasks and functions that users need to perform. For example, you may have roles such as "administrator," "manager," "editor," and "viewer," each with a distinct set of permissions. Role management also includes updating roles as needed, such as when new features are added to the application or when existing permissions need to be adjusted.

Role assignment is the process of associating users with specific roles. By assigning roles to users, you can ensure that each user has the appropriate level of access to perform their job duties without granting them unnecessary permissions. Role assignment can be done manually, through an automated process, or via integration with external identity providers, such as Azure AD or OAuth2.

To optimize role management and assignment, consider the following best practices:

- Define roles based on POLP, which means granting users the minimum permissions required to perform their tasks.

- Regularly review and update roles to ensure they accurately reflect the current requirements of your application.

- Implement a consistent process for role assignment, such as using templates or automation, to minimize human errors and simplify access management.

- Monitor role assignments and access logs to identify any discrepancies or potential security risks.

By effectively managing roles and assigning them to users, you can achieve a more secure and well-organized access control system within your SaaS application. This not only enhances security but also promotes a clear separation of responsibilities and facilitates compliance with data protection and privacy regulations.

Permission management and fine-grained access control

Permission management involves defining a set of actions or resources that users can access within your application. These permissions can then be assigned to roles or, in some cases, directly to users. Fine-grained access control goes beyond defining a set of actions or resources by allowing you to create highly detailed and specific permissions for a wide range of scenarios.

Fine-grained access control offers several benefits, including enhanced security, improved efficiency, and easier compliance. By providing users with only the necessary permissions, you minimize the potential for unauthorized access or actions that could compromise your application's security. With more precise access control, users can quickly find and interact with the resources they need while avoiding unnecessary clutter and distractions. For example, a marketing manager may only need access to customer data relevant to their campaigns and can avoid being overwhelmed by irrelevant data. If we think back to our secure building example, we can imagine that the areas of the building are very clearly signposted or maybe color-coded, making it very clear and obvious who is allowed into which parts of the building!

To implement fine-grained access control in your SaaS application, it's important to identify and define the specific actions and resources that users may need to access your application, taking into account different roles and responsibilities. Create a permissions hierarchy that organizes permissions logically and makes it easier to manage and maintain access control. Assign permissions to roles or users based on POLP, ensuring that users have the minimum access required to perform their tasks. For example, a customer support representative may only need access to customer records and basic account information, while a manager may need access to more sensitive financial data.

Regularly review and update permissions to ensure they accurately reflect the current requirements and functionality of your application. Monitor and audit permission assignments and access logs to detect discrepancies or potential security risks.

Summary

In this chapter, we explored the fundamental concepts of authentication and authorization in SaaS applications. Authentication is the process of verifying the identity of a user, typically through the use of credentials such as a username and password. Authorization, on the other hand, is the process of determining what actions a user is authorized to perform within the application, typically using **access control lists** (**ACLs**) or RBAC systems.

We discussed how authentication and authorization are closely related and work together to provide a secure environment for users to interact with the application. In a SaaS application, the consequences of a data leak can be severe, and getting authentication and authorization right is critical to preventing data leaks and protecting sensitive data.

We also discussed how implementing strong authentication and authorization mechanisms is particularly important in a multi-tenant application, where each tenant's data and resources must be protected from unauthorized access by other tenants or outside parties. Technical considerations for implementing authentication and authorization in a SaaS application include using a microservices architecture, implementing isolation techniques, and implementing automated testing and monitoring.

We explored some of the business considerations for implementing authentication and authorization in a SaaS application. These include clearly defining tenant boundaries and responsibilities, developing a clear pricing model, and providing a comprehensive onboarding (and offboarding) process.

Finally, we have worked through a practice example that adds authentication and authorization to our demo application!

By addressing both technical and business considerations, a SaaS application can provide a secure, reliable, and scalable platform that meets the needs of both application developers and tenants. Implementing strong authentication and authorization mechanisms can help to prevent data leaks and protect sensitive data while providing a clear and transparent pricing model and a comprehensive onboarding process can help to establish the application as a trusted provider of valuable services.

In the next chapter, we will learn about testing. Testing is a very important topic, particularly when dealing with SaaS applications. We will cover strategies for testing across the application stack.

Further reading

- Authentication with client-side Blazor using WebAPI and ASP.NET Core Identity: `https://chrissainty.com/securing-your-blazor-apps-authentication-with-clientside-blazor-using-webapi-aspnet-core-identity/`

- Blazor WebAssembly - User Registration and Login Example & Tutorial: `https://jasonwatmore.com/post/2020/11/09/blazor-webassembly-user-registration-and-login-example-tutorial`

- Introduction to Identity on ASP.NET Core: `https://learn.microsoft.com/en-us/aspnet/core/security/authentication/identity?view=aspnetcore-7.0&tabs=visual-studio&viewFallbackFrom=aspnetcore-2.2`

- Choosing A Master Password: `https://medium.com/edgefund/choosing-a-master-password-5d585b2ba568`

Questions

1. What is the difference between authentication and authorization in a SaaS application, and why are both important?

2. What are some of the technical considerations for implementing authentication and authorization in a SaaS application, and how can these help to prevent data leaks?

3. Why is implementing strong authentication and authorization mechanisms particularly important in a multi-tenant application, and what are some of the risks associated with not doing so?

4. What are some of the key business considerations for implementing authentication and authorization in a SaaS application, and how can these help establish the application as a trusted provider of valuable services?

5. What are some of the potential consequences of a data leak in a SaaS application, and how can implementing strong authentication and authorization mechanisms help to mitigate these risks?

6. How can automation be used to enhance the security of a SaaS application, and what are some of the benefits of doing so?

Part 4:
Deploying and
Maintaining the Application

This section focuses on what to do after the application has been built and how to keep it running smoothly in production as the user base starts to grow. As well as covering testing, this section covers monitoring and logging, **continuous integration/continuous deployment (CI/CD)**, and also offers advice on how to scale your SaaS application as the user base starts to grow.

This section has the following chapters:

- *Chapter 9, Testing Strategies for SaaS Applications*
- *Chapter 10, Monitoring and Logging*
- *Chapter 11, Release Often, Release Early*
- *Chapter 12, Growing Pains – Operating at Scale*

9

Testing Strategies for SaaS Applications

Testing is ubiquitous in the software industry, but often the reason for spending time doing the testing is lost. As we discuss the various testing techniques in this chapter, we will place a strong emphasis on the rationale behind each testing approach. By understanding not just the how but also the why of implementing these testing practices, you will be better equipped to make informed decisions about your testing strategy and ensure the long-term success of your **Software-as-a-Service (SaaS)** applications.

In this chapter, we will explore the important role that testing plays in the development and maintenance of SaaS applications. We'll use a combination of theory and practical examples to build a comprehensive understanding of the various testing approaches and their benefits. By the end of this chapter, you should have a solid foundation in testing strategies that will help you ensure the reliability, functionality, and overall quality of your SaaS applications.

We will start with a look at testing in general. This will include looking at the testing pyramid, a concept that illustrates the different types of testing—unit, integration, and **End-to-End (E2E)** testing—and their respective roles in the development process. This will give you a clear idea of the various testing approaches and their importance in ensuring that your application works as expected and meets the needs of your users.

Next, we will delve into **Test-Driven Development (TDD)**, a development methodology that emphasizes writing tests before writing the actual code. TDD has gained popularity in recent years due to its numerous benefits, such as improved code quality, faster development cycles, and easier maintenance. We will discuss the principles behind TDD and provide examples of how to apply them in your own projects.

We'll then go through the three broad categories of testing that are shown on the testing pyramid in a bit more detail and will look at how to apply these techniques to SaaS applications.

Throughout the chapter, we will cover the testing tools and frameworks commonly used in the Microsoft ecosystem. Understanding these tools will enable you to choose the most appropriate ones for your specific testing needs and help you create a more robust testing strategy for your SaaS applications.

Testing is a huge topic, and this chapter will only provide a general overview of the subject. However, by the end of this chapter, you should have a comprehensive understanding of how to approach testing for a SaaS application and an understanding of the various tools and techniques that are available.

The main topics covered in this chapter will be the following:

- Testing strategies that are specifically applicable to SaaS applications
- Test-driven development (TDD)
- End-to-endpyramid – unit, integration, and E2E testing

Technical requirements

All code from this chapter can be found at `https://github.com/PacktPublishing/ Building-Modern-SaaS-Applications-with-C-and-.NET/tree/main/Chapter-9`.

Testing strategies for SaaS applications

Testing is a fundamental aspect of the software development process. It helps to ensure the quality, reliability, and functionality of applications. Well-structured tests and test strategies allow developers to identify and fix issues early in the development process, which prevents costly and time-consuming errors that may surface later on.

As well as confirming the software is as bug-free as possible, testing also gives a way to verify that the software meets its requirements and performs as expected in a variety of scenarios. By incorporating testing practices into every step of the development process, developers can create more robust and maintainable applications, leading to improved user satisfaction and increased trust in the software, which ultimately improves the chances of the project being a success.

The importance of testing for SaaS applications

Insufficient testing can have serious consequences for any software application, including increased development costs, delayed releases, poor user experience, and reputational damage. When testing is inadequate, issues and defects are more likely to go unnoticed, leading to a higher likelihood of problems surfacing after deployment. This can result in time-consuming and costly fixes, as well as eroding user trust and satisfaction.

If the testing process is insufficient, then your users become your **Quality Assurance (QA)** team. And typically, users do not appreciate this!

It's important when developing any software application that your testing and QA are done before the application is in the hands of the users. This is doubly so with SaaS applications. These applications often serve multiple customers simultaneously, and any bug will affect all users simultaneously. Worse, a bug on one user instance can cause site-wide outages. Downtime or functionality issues can have a significant impact on user satisfaction, leading to customer churn and reputational damage.

SaaS applications typically require frequent updates and feature additions to stay competitive and meet the evolving needs of customers. A robust testing strategy allows developers to confidently release new features and updates without compromising the application's stability or introducing unforeseen issues. Finally, SaaS applications often involve complex interactions between various components, services, and APIs, making it essential to thoroughly test these interactions to ensure seamless operation and data integrity.

We'll start by looking at some best practices for testing your applications.

Testing best practices

Testing can be a challenging endeavor, but the benefits of getting it right are numerous, including improved code quality, increased confidence in your application's functionality, and reduced risk of defects reaching production. By following best practices, you can create a more robust and reliable testing process that not only uncovers issues early but also guides the design and development of your software. Throughout this section, we will provide a range of pointers and techniques to help you maximize the effectiveness of your testing efforts, ensuring that you can deliver high-quality software that meets the needs of your users.

- **Write testable code**: If you make the code easy to test, then the testing will be... easy! Follow SOLID principles, use dependency injection, create modular and decoupled components, and keep your classes small and well encapsulated. This is good advice in general, but it makes a huge difference to the testing process.

- **Test early, and test a lot**: The sooner in the development process that you start testing, the easier the process will be. Achieving 100% code coverage with tests is not really necessary, but achieving a high coverage percentage will generally result in better code, and fewer regressions. There are very few code bases that suffer from having too many tests, but there are many that suffer from having too few.

- **Maintain test isolation**: Each (unit) test should test only a single piece of the system, and should never depend on the results from any of the other tests. Integration and (E2E) tests may require more than a single piece of the system, but they should test only a single point of integration, or user interaction.

- **Keep tests simple and focused**: Each tests should be as short and as concise as possible. Tests should be easy to understand, and easy to maintain.

- **Use descriptive test names**: Name your tests in a way that clearly communicates their purpose and the expected outcome. It can be useful to start the name of the test with the word 'should' to describe what the object under the test is expected to do. `ShouldCorrectlyAddUpTheNumbers()` is a good name for a test that ensures numbers are added up correctly!

- **Avoid testing implementation details**: Focus on testing the behavior and functionality of your code, rather than its internal implementation. Try to test for a set of inputs to a function,; a particular output is generated. For example, if you're testing a function that calculates the sum of two numbers, focus on ensuring that the function returns the correct result for various input combinations rather than examining how the function performs the calculation internally. By doing so, you ensure that your tests remain relevant and useful, even if the implementation changes, as long as the expected behavior of the function stays consistent.

- **Foster a testing culture**: Encouraging a culture of testing within your team and organization is exceptionally important because it emphasizes the significance of testing in delivering high-quality software and encourages everyone to take responsibility for the overall quality of the product. A strong testing culture creates an environment where developers, testers, and other stakeholders actively collaborate to identify, prevent, and fix defects throughout the development process.

The next best practice is to use TDD. This warrants a subsection on its own!

Test-driven development (TDD)

TDD is a software development methodology that, at first glance, may seem counter-intuitive, as it emphasizes writing tests before writing the actual code. However, this approach has several benefits and helps developers create more robust, reliable, and maintainable software.

The core idea behind TDD is to create a failing test for a specific feature or functionality and then implement the code necessary to make the test pass. By writing the test first, developers are forced to clearly define the desired outcome and requirements for the code, which, in turn, leads to better overall design and structure. This process also helps developers catch any issues early in the development cycle, minimizing the likelihood of introducing errors or unexpected behavior.

Once the test has been written and the code has been implemented to pass the test, developers often refactor the code to improve its structure, readability, or performance. During this refactoring process, the existing tests serve as a safety net, ensuring that any changes made to the code do not break its functionality. This cycle of writing a test, implementing the code, and then refactoring as needed is repeated until the desired functionality is achieved. This is called the Red-Green-Refactor cycle.

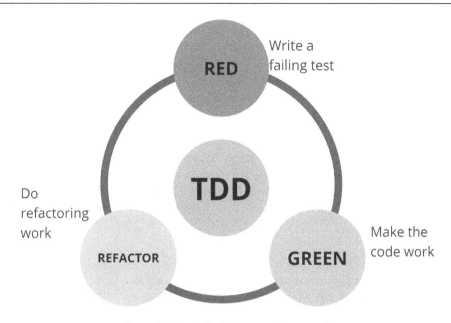

Figure 9.1 -- The Red -Green -Refactor cycle

Adopting TDD in a project can lead to several benefits. First, it promotes a more disciplined approach to coding, as developers must think about the requirements and desired outcomes before diving into implementation. Second, TDD simplifies debugging and maintenance, as the comprehensive test suite can quickly pinpoint issues and ensure that changes do not introduce new problems. Finally, TDD encourages better collaboration between team members, as the tests serve as clear documentation of the code's functionality and expected behaviour.

Types of TDD

TDD provides a general approach to writing tests before writing code, but there are also subtypes or variations of TDD that emphasize specific aspects or techniques. Some of these subtypes are listed here:

- **Behavior-Driven Development (BDD)**: BDD is an extension of TDD that focuses on the behavior of the software from the perspective of end-users or stakeholders. BDD encourages the use of a shared language and specification format (e.g., Gherkin) to describe the expected behavior of the software in a human-readable and easily understandable format. This shared understanding helps drive the creation of TDD tests, fostering better collaboration between developers, testers, and business stakeholders.

- **Acceptance Test-Driven Development (ATDD)**: ATDD is another variation of TDD that focuses on defining and validating acceptance criteria before starting the implementation of a feature. In ATDD, developers, testers, and business stakeholders collaborate to create acceptance tests that define the expected behavior of the system from a user's perspective. These tests are then used to guide the development process, ensuring that the resulting software meets the defined acceptance criteria.

- **Data-Driven Development (DDD)**: Not to be confused with Domain-Driven Design, Data-Driven Development in the context of TDD is an approach that focuses on using data to guide the creation of tests and the development process. Developers create test cases based on a range of input data and expected outcomes, ensuring that the code can handle various scenarios and edge cases. This approach is particularly useful when working with complex algorithms or data processing tasks.

- **Specification by Example (SBE)**: SBE is a collaborative approach to TDD that involves creating executable specifications based on real-world examples. Developers, testers, and business stakeholders work together to identify key examples that illustrate the desired behavior of the system. These examples are then used to create tests that guide the development process, ensuring that the resulting software meets the agreed-upon expectations.

These subtypes of TDD offer different perspectives and techniques for approaching test-first development.

Criticisms of TDD

While TDD has gained popularity and has many proponents, it has also faced criticism for various reasons. Some of the common criticisms of TDD include the following:

- **Overemphasis on testing**: Critics argue that TDD can lead to an excessive focus on writing tests at the expense of other important development tasks, such as architecture and design. This overemphasis on testing may result in developers spending too much time on writing tests and not enough on other aspects of the development process.

- **Incomplete test coverage**: TDD does not guarantee complete test coverage, as developers might not be able to anticipate all possible scenarios or edge cases while writing tests. This could lead to a false sense of security and potentially undetected bugs in the software.

- **Slow development process**: Writing tests before implementing the code can slow down the development process, especially for developers who are new to TDD. The additional time spent on writing and maintaining tests may be seen as an overhead cost that detracts from the overall development velocity.

- **Focus on unit tests**: TDD often leads to a focus on unit tests at the expense of other testing techniques, such as integration or E2E tests. While unit tests are valuable, they cannot catch all types of issues or verify the overall system's behavior, potentially leading to overlooked bugs or integration issues.

- **Overengineering**: TDD might encourage overengineering, as developers may be tempted to write code that satisfies the tests rather than focusing on the simplest and most efficient solution to the problem. This can lead to unnecessarily complex code that is harder to maintain and understand.

- **Learning curve**: TDD has a learning curve, and developers new to this approach may find it challenging to adapt their development process. They may struggle with writing effective tests, organizing their code, and following the red-green-refactor cycle, which can lead to frustration and decreased productivity.

Despite these criticisms, many developers and teams find TDD to be a valuable methodology that improves code quality, maintainability, and overall software reliability. The key to success with TDD is understanding its limitations and adapting the approach to suit the specific needs and constraints of a project. It is the opinion of the author of this book that TDD is an incredibly valuable part of the software development process – if it is done correctly.

Testing techniques

In the world of software testing, various techniques are employed to create effective and maintainable tests. These techniques help ensure that your tests are focused on the right aspects of your code, making it easier to identify and address potential issues. Employing appropriate testing techniques can lead to more reliable software, faster development cycles, and reduced maintenance efforts. By understanding and applying these techniques, you can create tests that are not only efficient but also easier to understand and maintain for your entire team.

Mocking

Mocking is a technique used in testing to replace real objects or services with simulated versions, known as mocks. The primary purpose of mocking is to isolate the code under test from its dependencies, enabling you to test individual components in isolation without relying on external factors. Mocks help you control the behavior of dependencies and verify that your code interacts correctly with them.

Common use cases for mocking include simulating the behavior of external services, such as APIs, databases, or third-party libraries, that may be unreliable, slow, or difficult to set up in a testing environment. By using mocks, you can focus on testing your own code's logic without worrying about the behavior of these external dependencies.

There are several popular mocking libraries for .NET, such as Moq, which simplifies the process of creating and managing mock objects in your tests. Moq allows you to create mocks of interfaces or abstract classes and define their behavior using a fluent API.

Stubbing

Stubbing is another technique used in testing, where you create lightweight objects called stubs that return pre-determined responses for specific method calls. Stubs are typically used for objects that are only used for retrieving data and don't need to have any complex logic or behavior. The main purpose of stubbing is to provide predictable and consistent test data, allowing you to focus on testing the code that consumes the data.

Here's a simple example of stubbing:

```
public class CustomerControllerb : ICustomerController
{
    public Customer GetCustomerById(int id)
    {
```

```
        return new Customer { Id = id, Name = "Dave
            Gilmore" };
    }
}
```

In the preceding snippet, a `Customer` stub is created with some predefined properties set.

Fakes

Fakes are simplified or partial implementations of classes or interfaces, used for testing purposes. They usually implement the same interface as the real object but provide a controlled environment for testing. Fakes can be hand-written or generated using a testing library. They can be used as a lightweight alternative to mocks and stubs when you need to simulate the behavior of a dependency without the complexity of a full implementation.

Fakes, stubs, and mocks are all conceptually quite similar, and can somewhat be used interchangeably depending on the exact details of the testing being performed.

The testing pyramid – unit, integration, and E2E testing

The testing pyramid is a concept that illustrates the optimal distribution of test types in a software project. It provides a visual representation of the relationship between unit, integration, and E2E testing, highlighting their relative importance and execution speed. Refer to the following diagram to better understand the structure of the testing pyramid:

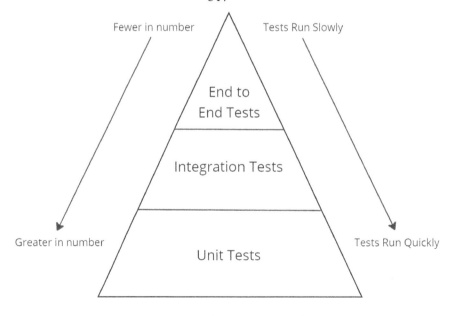

Figure 9.2 – The testing pyramid

At the base of the pyramid, we have unit tests. These tests are the most numerous and focus on verifying the correctness of individual components or functions in isolation. Unit tests are fast to execute, which enables developers to run them frequently during the development process.

In the middle of the pyramid, we find integration tests. These tests are fewer in number compared to unit tests, but they serve a vital purpose in validating the interactions between different components and services within the application. Integration tests take longer to run than unit tests, as they often involve more complex scenarios and dependencies.

At the top of the pyramid, we have E2E tests. These tests are the least numerous but are essential in ensuring the overall functionality and user experience of the application. E2E tests simulate real user scenarios by interacting with the application from start to finish, often through browser automation. As a result, they are slower to execute compared to unit and integration tests.

The testing pyramid emphasizes the importance of having a balanced testing strategy, with a larger number of fast unit tests, a smaller number of integration tests, and a few carefully chosen E2E tests. By understanding the role of each test type and their relative execution speeds, you can create an efficient and effective testing strategy for your SaaS applications.

Unit testing

Unit testing is the process of testing individual units or components of a software application in isolation. The primary goal of unit testing is to verify the correctness and reliability of each piece of code, ensuring that it functions as intended. By focusing on testing each component in isolation, developers can identify and fix issues at the earliest stages of the development process.

Improved code quality is one of the main benefits of unit testing. It encourages developers to write well-structured and modular code, leading to more maintainable and less error-prone applications. Unit testing also contributes to faster development, as it can catch issues early, minimizing the time spent on debugging and fixing issues. Additionally, unit tests serve as valuable documentation, providing insights into the intended behavior and functionality of each component.

Writing testable code using SOLID principles

To effectively leverage unit testing, it is essential to write testable code. Testable code is modular, with well-defined responsibilities for each component, making it easier to isolate and test individual units. One way to ensure that your code is testable is to follow the SOLID principles, which are a set of design guidelines aimed at promoting maintainability, flexibility, and testability in software development. The SOLID principles include the following:

- **Single Responsibility Principle (SRP)**: Each class or module should have a single responsibility or reason to change, ensuring that components have a focused purpose and are less likely to be affected by changes in other parts of the system.

- **Open/Closed Principle (OCP):** Software entities should be open for extension but closed for modification, meaning that existing code should not be altered when adding new functionality, thus reducing the risk of introducing bugs.

- **Liskov Substitution Principle (LSP):** Subtypes should be substitutable for their base types, ensuring that derived classes maintain the behavior of their base classes and do not introduce unexpected side effects.

- **Interface Segregation Principle (ISP):** Clients should not be forced to depend on interfaces they do not use. By creating small, focused interfaces, developers can avoid unnecessary dependencies and improve modularity.

- **Dependency Inversion Principle (DIP):** High-level modules should not depend on low-level modules but should depend on abstractions. This principle encourages the use of interfaces and abstract classes to decouple components, making it easier to test them in isolation.

Following the SOLID principles can help developers create code that is easier to test and maintain, ultimately improving the overall quality of the application.

TDD with unit tests

As mentioned earlier, TDD is a development methodology that emphasizes writing tests before writing the actual code. Unit tests play a crucial role in TDD, as they allow developers to verify the correctness of individual components and drive the implementation of new features.

In TDD, developers start by writing a failing unit test for specific functionality. The test should clearly define the desired outcome and requirements for the code. Next, the developer writes the minimum code required to make the test pass. This process ensures that each piece of code is written with a clear purpose and that its functionality is thoroughly tested.

Once the test passes, developers can refactor the code to improve its structure, readability, or performance while ensuring that the test still passes. This cycle of writing a test, implementing the code, and then refactoring as needed is repeated until the desired functionality is achieved. By using TDD with unit tests, developers can create more reliable, maintainable, and robust software applications.

The challenges and limitations of unit testing

While unit testing is probably the most conceptually straightforward of the three, it still has its own set of challenges and limitations. While unit tests are generally faster and more reliable than integration tests, they are limited by the scope of the code being tested. Unit tests focus on individual components in isolation, so they cannot detect issues that arise from interactions between components. This means that passing unit tests may not guarantee that the system will function correctly when integrated. Another challenge in unit testing is writing testable code, which requires following best practices such as SOLID principles and dependency injection. Properly mocking and stubbing dependencies can also be a challenge, as it may require a deep understanding of the dependencies' behavior to create accurate test doubles. Finally, unit tests can become brittle if they are too tightly coupled to the implementation details of the code, making it difficult to refactor the code without breaking the tests.

Integration testing

Integration testing is a vital part of the software development process that focuses on verifying the correct interaction between various components or modules within an application. As software systems grow more complex, the importance of ensuring that these interconnected pieces function together seamlessly becomes even more critical. In this section, we will discuss the essential aspects of integration testing, including testing API endpoints and working with databases. By understanding and implementing effective integration testing strategies, developers can build more reliable and robust software applications.

What integration testing is and why it matters

Integration testing is the process of verifying that the various components or modules of a software application work together correctly. Unlike unit testing, which focuses on testing individual components in isolation, integration testing aims to ensure that the components function as expected when integrated with one another. This is especially important in complex systems, where the interactions between components can lead to unexpected issues or failures.

Integration testing matters because it helps developers identify and fix problems that arise from the interactions between components. These issues may not be apparent during unit testing, as they only become evident when the individual components are combined. By performing integration testing, developers can ensure that the software functions correctly and reliably as a whole, providing a better overall user experience.

Testing API endpoints

API endpoints are a critical part of modern software applications, as they facilitate communication between different components or services. Integration testing of API endpoints involves verifying that the APIs return the expected results and behave correctly when called by other components in the system.

To test API endpoints, developers typically use tools such as Postman, Insomnia, or custom test scripts that send HTTP requests to the API and validate the responses. These tests can verify various aspects of the API, such as the following:

- **Response status codes**: This means ensuring that the API returns the expected status codes (e.g., 200 OK, 404 Not Found) for different scenarios
- **Response data**: This means verifying that the API returns the correct data in the expected format (e.g., JSON, XML)
- **Error handling**: This means checking that the API handles errors gracefully and returns meaningful error messages
- **Performance and reliability**: This means testing the API's performance under different loads and ensuring that it meets the required performance criteria

Integration testing with databases

Databases play a central role in many software applications, as they store and manage the data used by the system. Integration testing with databases involves verifying that the application interacts correctly with the database and ensuring that data is read, written, updated, and deleted as expected.

It is worth noting that testing databases can be challenging and is often skipped in favor of more robust testing around the application's interactions with the database. However, it is still good practice to try and test as much of the application as possible, so here are some pointers should you decide to go that route.

To perform integration testing with databases, developers can use various techniques, such as the following:

- **Using test data**: Developers can create test datasets that represent different scenarios, such as typical user data, edge cases, or invalid data. These datasets can be used to test the application's interaction with the database and validate that the data is processed correctly.

- **Mocking or stubbing database connections**: To isolate the application from the actual database during testing, developers can use mocking or stubbing techniques to simulate the database's behavior. This allows them to test the application's interaction with the database without actually connecting to it, making the tests faster and more reliable.

- **Testing database migrations**: In applications that use database migrations to manage schema changes, developers can test the migration scripts to ensure that they apply the changes correctly and do not introduce issues or data loss.

By performing integration testing with databases, developers can ensure that their application interacts correctly with the database and that the data is processed and stored reliably, providing a solid foundation for the overall functionality of the software.

The challenges and limitations of integration testing

Challenges and limitations of integration testing mainly arise from the increased complexity of interactions between components in a system. Integration tests often require more time and resources to set up, execute, and maintain due to the dependencies involved. The need to create test environments that closely resemble production can be both time-consuming and costly. Furthermore, integration tests can be less reliable, as they are more susceptible to issues caused by external factors such as network latency or third-party service outages. Additionally, integration tests tend to have a broader scope and are more complex, making pinpointing the root cause of a failure more difficult, leading to increased debugging time.

E2E testing

E2E testing is a crucial aspect of the software testing process that involves testing the entire application flow from the user's perspective. This type of testing verifies that all the components of an application work together seamlessly, ensuring that the application meets its intended functionality and provides a smooth user experience. E2E testing helps to identify any issues that may arise from the interaction between various components, which might not be detectable during unit or integration testing.

Encoding user journeys

E2E testing involves encoding real-life user journeys or workflows into test cases that simulate the user's interaction with the application. These user journeys cover the complete flow of the application, from the initial user input to the final output or result. By simulating user journeys, E2E testing ensures that the application behaves as expected and that any issues that may arise during real-world usage are detected and resolved before deployment.

Designing effective E2E test scenarios

Creating effective e2e test scenarios requires careful consideration of various factors. Developers should focus on identifying the most important and frequently used workflows or features of the application, as well as covering edge cases and potential failure points. Test scenarios should include uncommon or exceptional situations that may reveal hidden issues. Prioritization of test scenarios based on their importance, complexity, and potential impact on the application's overall functionality is also essential. Lastly, ensuring test maintainability is important—test scenarios should be easy to understand, update, and maintain as the application evolves.

The challenges and limitations of E2E testing

While E2E testing is a vital part of the software development process, it also comes with certain challenges and limitations. E2E tests can be time-consuming and resource-intensive, particularly when simulating complex user journeys or testing large applications. Test flakiness can sometimes occur due to factors such as network latency, timeouts, or unpredictable user behavior, leading to inconsistent results. As the application evolves, E2E tests may need frequent updates to reflect changes in the application's features and workflows, which can make test maintenance more challenging. Additionally, it may not be feasible to cover all possible user journeys and scenarios in E2E tests, which could result in undetected issues.

Despite these challenges, E2E testing remains a critical component of the software testing process, helping to ensure that applications function correctly and provide a reliable user experience. By designing effective E2E test scenarios and addressing the challenges and limitations, developers can build high-quality, robust applications.

An overview of testing tools and frameworks for SaaS applications

There are a huge number of tools and frameworks available for running tests, each with its own strengths and weaknesses. However, in this section, we will limit our focus to those that are applicable to Microsoft technologies, such as those we have used in the demo application. By narrowing our scope, we can provide a more targeted and relevant discussion for developers working with these technologies in their SaaS applications.

General testing of .NET applications

When developing SaaS applications (or any applications!) using .NET, it is important to ensure that the code is well-tested and reliable. Two of the most popular testing frameworks for .NET are xUnit and NUnit. Both of these frameworks are open source, widely used, and well-supported by the .NET community. They provide a rich set of features and functionality that enable developers to write comprehensive tests for their applications.

xUnit is a modern and extensible testing framework specifically designed for .NET. It is the default testing framework for .NET Core and ASP.NET Core projects, making it an excellent choice for developers working on modern .NET applications. Some of its key features include the following:

- A clean and simple syntax for writing tests
- Support for parallel test execution, which can speed up the testing process
- A robust and flexible set of assertions and test attributes

NUnit is another popular testing framework for .NET, with a long history of use in the .NET community. Although it is not the default testing framework for .NET Core and ASP.NET Core projects, it still enjoys widespread support and provides a solid set of features for writing unit tests. Some key features of NUnit include the following:

- A familiar syntax for writing tests, particularly for developers with experience in other testing frameworks
- Support for parallel test execution
- A comprehensive set of assertions and test attributes

There is really very little difference between the two, and the choice of which to use will largely come down to individual preference and will not impact your project much at all.

In addition to xUnit and NUnit, there are other useful tools and libraries that can be employed for testing .NET applications, such as:

- **Moq**: This is a popular mocking library for .NET, which can be used to create mock objects and set up expectations for their behavior in tests.

- **FluentAssertions**: This is a library that provides a more readable and expressive syntax for writing assertions in tests, making it easier to understand the intent of the test.

- **NSubstitute**: An alternative to Moq, NSubstitute is another popular mocking library for .NET. It provides a simple and intuitive syntax for creating mock objects and defining their behavior in tests. NSubstitute can be used with NUnit, xUnit, and other testing frameworks.

- **AutoFixture**: AutoFixture is a library that helps automate the generation of test data for unit tests. It can create objects with random or customized values, making it easier to set up test scenarios with minimal manual configuration. AutoFixture can be used in conjunction with other testing frameworks such as NUnit and xUnit.

- **Shouldly**: Shouldly is an assertion library similar to FluentAssertions that aims to provide a more human-readable and expressive syntax for writing assertions in tests. It simplifies the process of writing assertions and makes it easier to understand the intent of the test.

- **SpecFlow**: SpecFlow is a BDD tool for .NET that enables developers to write tests in a natural language format using Gherkin syntax. It allows non-technical stakeholders to understand and contribute to the test scenarios, bridging the gap between development and business teams.

Testing APIs

When it comes to writing automated tests for Web APIs, tools such as Postman and Newman can be invaluable. Postman is a powerful API testing tool that allows developers to send HTTP requests to API endpoints and inspect the responses, making it easier to debug and validate the API's behavior during development. Newman, on the other hand, is a command-line companion tool for Postman that allows you to run Postman collections directly from the command line or as part of your **Continuous Integration/Continous Deployment (CI/CD)** pipeline.

We have used Thunder Client throughout the examples in this book, primarily to keep everything contained inside **Visual Studio Code (VSCode)**. Postman offers a few more advanced features, such as pre-request scripts and document generation. As your SaaS project grows, there may be advantages in using Postman over Thunder Client. Thunder Client is a lightweight and easy-to-use option for developers who want a simple API testing tool integrated with their VSCode environment. Postman, on the other hand, is a more powerful and feature-rich tool suitable for advanced API testing scenarios and team collaboration. Your choice between the two will depend on your specific requirements and personal preferences.

It can be slightly tricky to mock an HTTP client when testing APIs, but there are libraries such as `Moq` and `HttpClient Interception` that can help simplify this process. API testing can also be considered a form of integration testing since it involves verifying the correct interaction between various components of the API.

Testing Blazor applications

Testing Blazor applications can be a bit more challenging due to the nature of the technology. However, there are several tools and libraries available to help make the process easier:

- **bUnit**: This is a testing library specifically designed for Blazor applications, allowing developers to write unit and component tests

- **Playwright**: This is a browser automation library that can be used to write E2E tests for Blazor applications, simulating user interactions and verifying the application's behavior

- **Selenium**: While not specifically designed for Blazor, Selenium is a popular browser automation tool that can also be used to write E2E tests for Blazor applications

The challenges of writing tests for databases

Testing database-related code can be challenging due to the inherent complexities of working with databases. It is relatively uncommon to write tests specifically for database interactions, but there are some reasons and general pointers to consider:

Databases can introduce statefulness and external dependencies into tests, making it harder to maintain isolated and deterministic test environments.

It may be more efficient to focus on testing the application's data access layer and business logic rather than directly testing the database itself.

When testing code that interacts with databases, consider using techniques such as mocking or stubbing to isolate the database-related code and simulate the expected behavior of the database.

To test database-related code more effectively, consider using a dedicated database testing tool such as tSQLt for SQL Server, which allows you to write unit tests for your database objects (such as stored procedures, functions, and triggers).

By considering these factors and employing the appropriate tools and techniques, you can improve the quality of your SaaS application through comprehensive testing across all aspects of the application.

A practical demonstration

While it is outside the scope of this book to provide a full suite of tests, we can add some unit tests to the GoodHabits app that we have built by way of a practical demonstration of some of the tools and techniques that we have discussed in this chapter.

Let's start by adding a test project. We will use xUnit, as it is generally recommended for modern .NET applications. We will also add Moq and Fluent Assertions to this project and have a look at how we can use them.

Do this with the following script:

```
mkdir test; \
cd test; \
dotnet new xunit --name GoodHabits.HabitService.Tests; \
cd GoodHabits.HabitService.Tests; \
dotnet add reference ../../GoodHabits.HabitService/
    GoodHabits.HabitService.csproj; \
dotnet add package Moq; \
dotnet add package FluentAssertions; \
rm UnitTest1.cs ; \
touch HabitsControllerTest.cs ; \
cd ../..
dotnet sln add ./tests/GoodHabits.HabitService.Tests/
    GoodHabits.HabitService.Tests.csproj;
```

The preceding script will add a file called `HabitsControllerTest.cs` that we will use to test `HabitsController`. Add the following code:

```
using Moq;
using FluentAssertions;
using GoodHabits.HabitService.Controllers;
using Microsoft.Extensions.Logging;
using GoodHabits.HabitService;
using AutoMapper;
using Microsoft.AspNetCore.Mvc;
public class HabitsControllerTests
{
    private readonly HabitsController _habitsController;
    private readonly Mock<ILogger<HabitsController>>
    _loggerMock;
    private readonly Mock<IHabitService> _habitServiceMock;
    private readonly Mock<IMapper> _mapperMock;

    public HabitsControllerTests()
    {
        _loggerMock = new Mock<Ilogger
            <HabitsController>>();
        _habitServiceMock = new Mock<IHabitService>();
        _mapperMock = new Mock<IMapper>();
        _habitsController = new HabitsController
            (_loggerMock.Object, _habitServiceMock.Object,
                _mapperMock.Object);
    }
```

```
[Fact]
public async Task GetVersion_ReturnsExpectedVersion()
{
    var result = await _habitsController.GetVersion();
    var okResult = result.Should().BeOfType
        <OkObjectResult>().Subject;
    okResult.Value.Should().Be("Response from version
        1.0");
}
}
```

You can now run the tests by opening a terminal and typing `dotnet test`. You should see the following indicating that the test has passed:

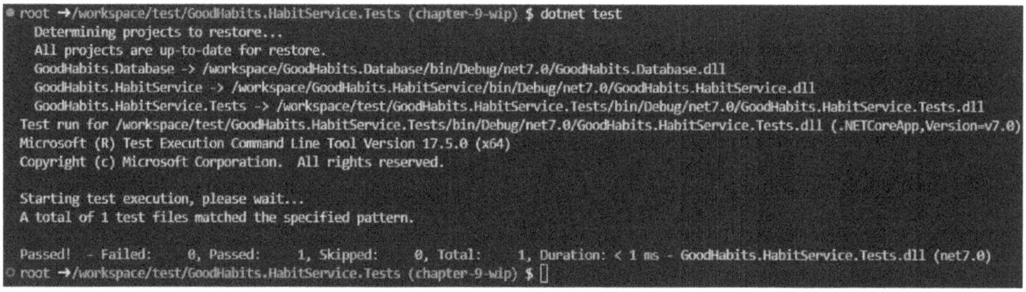

```
 root →/workspace/test/GoodHabits.HabitService.Tests (chapter-9-wip) $ dotnet test
  Determining projects to restore...
  All projects are up-to-date for restore.
  GoodHabits.Database -> /workspace/GoodHabits.Database/bin/Debug/net7.0/GoodHabits.Database.dll
  GoodHabits.HabitService -> /workspace/GoodHabits.HabitService/bin/Debug/net7.0/GoodHabits.HabitService.dll
  GoodHabits.HabitService.Tests -> /workspace/test/GoodHabits.HabitService.Tests/bin/Debug/net7.0/GoodHabits.HabitService.Tests.dll
Test run for /workspace/test/GoodHabits.HabitService.Tests/bin/Debug/net7.0/GoodHabits.HabitService.Tests.dll (.NETCoreApp,Version=v7.0)
Microsoft (R) Test Execution Command Line Tool Version 17.5.0 (x64)
Copyright (c) Microsoft Corporation.  All rights reserved.

Starting test execution, please wait...
A total of 1 test files matched the specified pattern.

Passed!  - Failed:     0, Passed:     1, Skipped:     0, Total:     1, Duration: < 1 ms - GoodHabits.HabitService.Tests.dll (net7.0)
 root →/workspace/test/GoodHabits.HabitService.Tests (chapter-9-wip) $ []
```

Figure 9.3 – The first test passed

The preceding test is an extremely simple test, to ensure that we get the correct string back from the version endpoint. But we have demonstrated some advanced testing techniques. We have used the `Moq` package to create mocks of all of the dependencies of the controller.

We have also used the `FluentAssertions` library to make the test very readable. The intent of this line should be very obvious just from reading it!

```
okResult.Value.Should().Be("Response from version 1.0")
```

This has been a very gentle introduction to testing—there is a lot more that could be done to prove the correct operation of the `HabitsController` class! It would be an excellent exercise to start building out this test suite and maybe add some tests for the other projects. Or even add some integration and E2E tests!

Summary

In this chapter, we have explored the important role that testing plays in the development and maintenance of SaaS applications. By understanding the various types of testing—unit, integration, and E2E testing—and their respective roles in the development process, you are now better equipped to implement a comprehensive testing strategy for your applications.

We have also discussed TDD and its benefits, such as improved code quality, faster development cycles, and easier maintenance. By incorporating TDD into your development process, you can further enhance the reliability and functionality of your SaaS applications.

We have also taken a high-level overview of testing by looking at some of the underlying techniques and the tools that you can use to apply those techniques.

This chapter has provided you with a comprehensive understanding of the important role that testing plays in SaaS application development. We hope that you can now confidently apply these concepts and practices to your own projects, resulting in more robust, reliable, and high-quality SaaS applications.

As you continue to develop and deploy your SaaS applications, it is essential to monitor their performance and log relevant information to ensure smooth operation and quickly address any issues that may arise.

In the next chapter, we will discuss monitoring and logging, covering the essential tools and best practices to help you maintain and optimize your SaaS applications in a production environment.

Further reading

- What Is Unit Testing? `https://smartbear.com/learn/automated-testing/what-is-unit-testing/`

- Integration Testing: What is, Types with Example: `https://www.guru99.com/integration-testing.html`

- Test Razor components in ASP.NET Core Blazor: `https://learn.microsoft.com/en-us/aspnet/core/blazor/test?view=aspnetcore-7.0&viewFallbackFrom=aspnetcore-7.0`

- What is Test Driven Development? TDD vs. BDD vs. SDD: `https://testrigor.com/blog/what-is-test-driven-development-tdd-vs-bdd-vs-sdd/`

- Unit Testing: Why Bother? `https://www.cmsdrupal.com/blog/unit-testing-why-bother`

Questions

1. What are the three main types of testing in the testing pyramid, and what is the primary purpose of each type?

2. How does TDD improve code quality, development speed, and maintainability?

3. What are some popular testing tools and frameworks for the Microsoft ecosystem, and what are their primary uses?

4. How can unit testing help ensure the correctness and reliability of individual components in a SaaS application?

5. Why is integration testing important in validating the interactions between different components and services in a SaaS application?

6. How does E2E testing contribute to ensuring the overall functionality and user experience of a SaaS application?

10

Monitoring and Logging

A typical **Software as a Service** (**SaaS**) application will cater to millions of users who access the platform around the clock. When unexpected issues arise, it can be incredibly difficult to diagnose, recreate, and resolve these problems. Monitoring and logging are essential tools that can address this challenge by providing invaluable insights into the health and performance of applications that are running in production environments and being used for real.

Monitoring focuses on proactively observing the system's health and performance by collecting and analyzing key metrics in real time. This is a "top-down" view of the overall health of the entire system, including things such as resource utilization. This process enables developers to identify potential issues, optimize resource utilization, and maintain a seamless user experience. Technologies such as Application Insights and Azure Monitor offer tailored solutions to effectively implement monitoring strategies in SaaS applications, ensuring reliability and high performance.

Conversely, logging is vital for capturing a wide range of events occurring within an application. Logging is a bit more fine-grained than monitoring, and will generally capture issues and events that have occurred in the code of the application. Detailed information about errors, user actions, and system events allows developers to effectively diagnose and troubleshoot issues while maintaining a comprehensive audit trail for security and compliance purposes. By leveraging logging libraries, developers can seamlessly integrate logging into their SaaS applications.

This chapter will cover the following topics:

- Monitoring
- Logging
- Monitoring and logging considerations for SaaS applications

This chapter delves into monitoring and logging within the context of SaaS applications, highlighting the unique challenges and considerations that arise. Practical guidance will be provided to help implement efficient monitoring and logging strategies, helping developers of SaaS applications to maintain a high-performing and reliable application.

Let's start with an overview of both, and then move on to looking at them each in more detail.

Overview

Both monitoring and logging are tools that you can use to see how your SaaS application is performing. Because a SaaS application typically has many moving parts, getting an overall view of the system health can be very tricky and can involve a number of different skill sets.

For example, if a user reports that the application is "slow," this could be caused by the following:

- The user's personal computer or network being slow, in which case there is nothing that we can do about it!

- The network connection to the cloud services being slow, in which case we need a network specialist to resolve it, and probably assistance from the network provider to increase the bandwidth.

- The API could be having trouble handling multiple concurrent requests, in which case we would need a backend developer to identify and DevOps specialists to correctly scale the API.

- The UI could be making very inefficient requests to the API, resulting in slow performance. This would require a coordinated effort from a frontend and a backend developer to resolve.

- The database could be the bottleneck. Perhaps the indexing on the database is insufficient, and so a **Database Administrator** (**DBA**) is required to identify and correct the issue.

I could continue! Diagnosing customer issues in SaaS applications can be extremely difficult and it can require a wide array of specialists to identify and resolve the problem.

When the app is running in production, the developers of the app have very little insight into any issues that arise in the application, and so typically, monitoring and logging techniques are used to keep track of what is happening in real time. Without these tools, diagnosing issues in production apps is little more than guesswork and is a hugely time-consuming and frustrating exercise.

As a very broad overview, we could say that both monitoring and logging give an insight into what is happening when an application is running in production. These insights are used by the developers and maintenance teams to more quickly diagnose and correct user issues as they arise.

However, that is a little high level, so we will dig in a bit deeper!

Monitoring generally focuses on the overall health and performance of the application and its components from an external perspective, including app services, networking, and databases. It provides a high-level view of the system's operation and identifies potential issues and bottlenecks. Monitoring is generally proactive and can be used to identify issues before they arise, such as a storage device running out of space or the available bandwidth starting to approach its limit. You can think of monitoring as a top-down process that is outside of the application looking in. Monitoring is outside looking in.

Logging, on the other hand, is more focused on capturing detailed information about the events, errors, and transactions occurring within the application code itself. This detailed data helps developers to diagnose, troubleshoot, and understand specific issues related to the application's inner workings. So, while monitoring offers a broader, outside-in perspective, logging delves into the finer aspects of the application code and records its behavior. Logging is always retroactive. It is a store of things that have already happened in the application. You can think of logging as a more fine-grained process that is backed up in the application, reporting events and actions to some external store of such actions. Logging is inside looking out.

The following table shows the difference between monitoring and logging:

Monitoring	Logging
Real-time observation of performance and resource usage	Recording of events, errors, and transactions
Focus on system health and availability	Focus on detailed information and audit trails
Proactive detection of anomalies and potential issues	Retrospective analysis of historical data
Optimization of resource utilization	Diagnosis and troubleshooting of application issues
High-level view of application components	In-depth understanding of application code behavior
Outside looking in	Inside looking out

Table 10.1 – Difference between monitoring and logging

You can see that while the two topics are related, there are quite significant differences between the function and purpose of each. We will now look at both in detail, starting with monitoring.

Monitoring

Monitoring is the process of continuously observing and measuring various aspects of a system, such as performance, resource utilization, and availability, to ensure its optimal functioning and identify potential issues. In the context of SaaS applications, monitoring involves the collection and analysis of key metrics and events in real time, allowing developers to proactively detect anomalies, optimize resources, and maintain a seamless and reliable user experience.

Monitoring is a critical aspect of maintaining the health and performance of SaaS applications. In an environment where millions of users access the platform 24 hours a day, performing a wide range of actions, proactive observation of the system becomes essential. This section will explore the key concepts, tools, and strategies for implementing effective monitoring in SaaS applications.

Key aspects of monitoring

There are a few key considerations to keep in mind when building a monitoring system for your application:

- Performance metrics are crucial for gauging the responsiveness and efficiency of a SaaS application. These metrics can include response times, throughput, error rates, and latency, among others. By closely monitoring these parameters, developers can identify bottlenecks and areas for optimization, ensuring a smooth and satisfying user experience.

- Resource utilization monitoring involves keeping track of how the application consumes system resources such as CPU, memory, disk space, and network bandwidth. By monitoring resource consumption, developers can detect and prevent issues related to resource contention or exhaustion, which can negatively impact the performance and stability of the application. This insight also aids in making informed decisions about scaling and infrastructure management. In the cloud-first world we now work in, resource utilization has a significant cost to the business, and so it is now even more important to have a handle on this at all times.

- Application availability and health monitoring focus on assessing the operational state of the application and its components. This includes monitoring the uptime, error rates, and performance of individual services or components within the system. By tracking the health of the application, developers can proactively detect and address issues before they escalate, minimizing downtime and maintaining a high level of service for users.

- Long-term trends and capacity planning involve analyzing historical monitoring data over a period of time to identify patterns and forecast future system requirements. By understanding trends in user growth, resource consumption, and performance metrics, developers can make informed decisions about infrastructure investments, optimize resources, and prepare the application for increased demand. This foresight enables SaaS providers to deliver a consistently reliable and performant service, even as user bases and workloads evolve over time.

If you keep these four key considerations in mind, you should be well on your way to delivering a successful monitoring system for your SaaS application. Of course, there is more to it than that! So, we'll now take a look at some nuances that you may encounter as a SaaS developer.

Monitoring tools

We have discussed the importance of monitoring your SaaS application. We will now look at the tools that you can use to perform this important task.

In general, it is advisable to use an "off-the-shelf" monitoring solution rather than trying to custom-build this functionality. The monitoring tool can be as complex as the application that it is monitoring! These tools provide a highly specialized function, which is generally better left to specialists to implement. There are many different options for monitoring, but in general, a good monitoring tool should provide the following functionalities:

- **Collect and display relevant data**: These are the absolute basics of monitoring! A good monitoring tool should be able to collect and display a wide range of relevant data, including server performance metrics, application-specific metrics, and user behavior data.

- **Provide real-time monitoring**: Real-time monitoring is crucial to quickly detect and respond to issues as they arise. A good monitoring tool should be able to provide real-time updates on the status and performance of your application, viewable via a dashboard or something similar.

- **Alert and notify**: The tool should be able to alert and notify you when issues are detected, via methods such as email, SMS, or chat tools such as Slack. It is not reasonable to expect a team member to keep an eye on the dashboard 24 hours a day, so an alert system can be used to inform the team that something has gone wrong. The tool should also provide customizable alert thresholds so that you can set the appropriate level of urgency for different types of issues. This is important as frequent non-important error messaging will result in people ignoring all of the messages, and so miss the important ones.

- **Enable proactive monitoring**: In addition to reacting to issues as they occur, a good monitoring tool should enable proactive monitoring by providing insights into potential issues before they impact your users. This can be achieved through features such as predictive analytics and trend analysis, allowing the team to act in advance to prevent issues before they arise.

- **Support customization**: No two SaaS applications are the same, so the tool should allow for a high level of customization and configuration to meet the specific needs of your application. This includes the ability to create custom dashboards and reports, as well as integrating them with other tools and systems – particularly logging systems, which we will cover later in this chapter.

- **Provide scalability and reliability**: A good monitoring tool should be able to handle large amounts of data and provide reliable performance even under high loads. It should also be able to scale up or down as needed to accommodate changes in your application's usage patterns.

- **Facilitate collaboration**: A good monitoring tool should facilitate collaboration between different teams and stakeholders involved in maintaining and improving your application. As an application grows, there will be multiple teams interested in the different aspects of the application's overall health. Each of these user classes should be able to get what it needs from the monitoring tool, by making use of features such as role-based access control and the ability to share dashboards and reports.

It is very hard to recommend a specific tool to use for monitoring, as the best choice will depend on the tech stack that is being implemented. Given that this book is focusing on .NET and the Microsoft tech stack, it is probably safe to say that Azure-based systems such as Application Insights or Azure Monitor will be the most useful.

Here is a list of some commonly used monitoring tools that you may wish to consider. Note that there is some overlap here with tools for logging, as we will see later in this chapter:

- **Application Insights**: A Microsoft Azure-based monitoring service that provides comprehensive application performance monitoring and diagnostics for .NET applications.

- **Azure Monitor**: A Microsoft Azure service for collecting, analyzing, and acting on telemetry data from various Azure and on-premises resources, including application and infrastructure monitoring.

- **Datadog**: A cloud-based monitoring and analytics platform that provides full stack observability across applications, infrastructure, and cloud services.

- **New Relic**: A comprehensive application performance monitoring and management platform, offering real-time visibility into the performance and health of applications and infrastructure.

- **Prometheus**: An open source monitoring and alerting toolkit, primarily designed for reliability and scalability, often used with container orchestration systems such as Kubernetes.

- **Grafana**: A popular open source visualization and analytics platform that allows users to create and share interactive dashboards and alerts using data from various monitoring tools.

- **Elasticsearch**, **Logstash**, **Kibana** (**ELK**) **Stack**: A popular open source log management and analytics platform that combines Elasticsearch for search and analytics, Logstash for log processing, and Kibana for data visualization.

How to do this

We have talked a lot about the tools that we could use for this, but not so much about how to actually do it! Here's a list of steps that you may want to consider when setting up a monitoring strategy. Remember, monitoring is "outside looking in":

1. Define the metrics that are important for your specific application. There is no one-size-fits-all here; you will need to look carefully at what information may come in handy for your application.

2. Pick a tool. Again, there is no one "best" tool to use. Study the options available to you and decide which one is most applicable. These tools are typically paid-for services, so create an invoice and purchase the tool.

3. Configure the monitoring tool to collect the defined metrics. Depending on the tool you chose, this may involve installing agents on your servers, configuring API integrations, or setting up custom scripts.

4. Set up appropriate thresholds, alerts, and notifications for the metrics you're monitoring. This will help you proactively detect anomalies, performance issues, or potential bottlenecks before they impact your users.

5. Integrate your monitoring tool with your existing development and operations workflows, such as your issue-tracking system, CI/CD pipelines, and communication platforms. This will ensure that your team is promptly informed of any issues and can take action immediately.

6. Continuously review and refine your monitoring strategy as your application evolves. As new features are added, performance requirements change, or user expectations grow, you may need to adjust your monitoring approach accordingly.

7. Regularly analyze the collected monitoring data to identify trends, patterns, and potential areas for optimization. This will help you make informed decisions about your application's architecture, resource allocation, and future development priorities.

By following these steps and tailoring your monitoring strategy to the unique needs of your SaaS application, you'll be well equipped to maintain a reliable, high-performing, and resilient platform for your users.

Best practices for monitoring

Monitoring can be challenging for any application, and the complexity increases significantly in the context of SaaS applications. In this section, we will explore a set of best practices tailored to effectively monitor SaaS applications to give you the best chance of success:

- **Defining relevant metrics and thresholds**: When monitoring SaaS applications built with Microsoft technologies, it's essential to define relevant metrics and thresholds that accurately represent the application's health and performance. This may include metrics such as response times, error rates, resource utilization, and throughput. Establishing appropriate thresholds for these metrics will help you identify potential issues before they escalate and affect the user experience.

- **Implementing proactive monitoring and alerting**: Proactive monitoring involves continuously observing your application's performance and health, allowing you to detect issues early and take corrective action. With Microsoft technologies, tools such as Application Insights and Azure Monitor can be used to set up proactive monitoring and alerting. By configuring alerts based on predefined thresholds, you can ensure that your team is notified of potential issues as soon as they arise, minimizing downtime and maintaining a high-quality user experience.

- **Ensuring data privacy and compliance in multi-tenant environments**: SaaS applications often serve multiple tenants within a single application instance, raising data privacy and compliance concerns. When monitoring multi-tenant applications, it's crucial to maintain appropriate data isolation and ensure that tenant-specific performance data is not accessible to other tenants. Microsoft technologies, such as Azure Monitor, can help you implement tenant-specific monitoring while adhering to privacy and compliance requirements.

- **Integrating monitoring data with logging and other diagnostic tools**: Monitoring and logging complement each other by providing different insights into your application's performance and health. Integrating monitoring data with logging and other diagnostic tools can help you gain a comprehensive understanding of your application's behavior and identify the root causes of issues more effectively. Tools such as Application Insights and Azure Monitor can be integrated with logging platforms such as ELK Stack or Azure Log Analytics, enabling you to correlate monitoring and log data for deeper analysis.

- **Alerting and notifications in monitoring**: In addition to collecting and analyzing monitoring data, it's crucial to establish an effective alerting and notification system for your SaaS application. Alerting involves configuring predefined thresholds for relevant metrics, and when these thresholds are breached, notifications are sent to the appropriate team members, allowing them to respond quickly and mitigate any potential impact on the user experience. Microsoft technologies, such as Application Insights and Azure Monitor, offer robust alerting capabilities that can be customized to match your application's unique needs. By integrating these alerting features with communication tools, such as email, SMS, or collaboration platforms such as Microsoft Teams or Slack, you can ensure that your team stays informed of any critical issues and can take timely action to resolve them.

- **Continuously refining and improving monitoring strategies**: Monitoring strategies should evolve alongside your application, as requirements and performance goals change over time. Continuously reviewing and refining your monitoring strategies ensures that you remain focused on the most relevant metrics and can proactively address emerging issues. By leveraging the insights and analytics provided by monitoring tools such as Application Insights and Azure Monitor, you can continuously improve your monitoring approach and maintain a high-performing, reliable SaaS application.

In this section, we have looked at the reason for monitoring in the first place, considered its applicability to SaaS applications, looked at the available tools, and considered the best practices. We'll now move on to think about logging.

Logging

Logging, in contrast to monitoring, focuses on capturing detailed information about the events, user actions, and system behavior within your application. While monitoring provides a high-level view of your application's performance and health, logging allows you to dive deeper into specific events and occurrences, enabling effective troubleshooting and maintaining a comprehensive audit trail for security and compliance purposes.

Logging is the practice of capturing and recording detailed information about events, errors, and user actions that occur within a system, providing developers with valuable insights for troubleshooting and diagnosing issues. While monitoring focuses on the real-time observation of system health and performance, logging is more concerned with maintaining a comprehensive record of application events and activities for future analysis.

Logging plays an indispensable role in maintaining and improving SaaS applications, as it allows developers to understand the intricate interactions and processes occurring within the application. With millions of users constantly interacting with the platform, having a detailed log of system events becomes crucial for identifying the root cause of issues and ensuring smooth operation. This section will delve into the key concepts, tools, and techniques for implementing effective logging in SaaS applications. By employing logging practices tailored to the unique requirements of SaaS environments, developers can enhance their ability to diagnose and resolve issues, maintain a robust audit trail for security and compliance purposes, and ultimately deliver a reliable and high-performing service to their users.

Key aspects of logging

In this section, we will examine the key aspects of logging that are essential for implementing a comprehensive and effective logging strategy in SaaS applications, enabling developers to gain valuable insights, maintain robust audit trails, and ensure optimal application performance.

The basis of any logging system is the ability to collect information from various sources, such as applications, databases, and a collection of microservices or containers. The ability to do this efficiently is the foundation of any successful logging strategy. A well-designed log collection system should be able to handle the diverse types and volumes of log data generated by your application while minimizing the impact on application performance. Ensure that all relevant log data is captured and available for analysis.

Once log data has been collected, it needs to be stored in a centralized and easily accessible location. Effective log storage strategies focus on data retention, ensuring that log data is preserved for an appropriate length of time and can be quickly retrieved when needed. Scalability is also a crucial consideration, as log storage systems must be able to grow to accommodate increasing data volumes as your SaaS application expands. Do not underestimate the amount of data that can be collected by the logging system! Plan accordingly, as the data can be very expensive to store in cloud infrastructure.

There is no point in gathering and storing the data if it is hard to get any usable information out of the logs. A system should be in place to allow relevant parties to read and analyze the log data to identify patterns, trends, and anomalies. This can help developers diagnose and troubleshoot issues more effectively, optimize resource utilization, and even identify potential security threats – hopefully before they have occurred! To facilitate quick insights and decision-making, it's important to present log data in an easily digestible format, such as as charts, graphs, and dashboards. Log visualization tools such as Kibana, Grafana, and Azure Monitor can help transform raw log data into meaningful visual representations, making it easier for developers and operations teams to understand the state of the application and identify areas for improvement. These tools can also be customized to create tailored visualizations that highlight the most relevant information for your specific SaaS application.

With the vast amounts of log data generated by SaaS applications, it's essential to filter out irrelevant or noisy log data and focus on the most critical and actionable information. Log-filtering techniques can be employed at various stages of the logging process, from collection to analysis, to help reduce noise and improve the signal-to-noise ratio. By implementing effective log-filtering strategies, developers can save time and resources by concentrating on the most pertinent log data and ensuring that important events do not get lost in the noise.

Ensuring the confidentiality, integrity, and availability of log data is a key aspect of logging, as it involves compliance with data protection regulations and adherence to industry best practices. Log security measures may include encryption, access control, and data backup strategies, all aimed at safeguarding log data from unauthorized access, tampering, or loss.

Configuring alerts based on specific log events or patterns is crucial for proactively identifying potential issues in your SaaS application. Log alerting enables timely notifications to be sent to the appropriate team members when potential issues are detected, allowing for swift action to be taken to resolve them.

Finally, it is not necessary to keep all of the log data forever, but it can be useful to retain some data for a longer term. Preserving some form of historical log data for future reference, analysis, or compliance purposes can be very useful and should be considered when building up your logging system.

Logging tools

In general, it is advisable to use off-the-shelf logging solutions. Logging is a quite mature and well-understood concept now, so there is generally little benefit to building your own custom implementation. In this section, we will look at some general pointers to help choose a good logging tool, and then consider some specific tools:

- **Collect and store logs**: A good logging tool should be able to collect and store logs from various sources, such as servers, applications, and databases. It should also be able to handle large volumes of logs and store them in a scalable and efficient manner.

- **Provide search and analysis capabilities**: A good logging tool should provide robust search and analysis capabilities that allow you to easily search and filter through logs to identify issues and troubleshoot problems. It should also support advanced querying and filtering to enable more complex analysis.

- **Enable real-time monitoring**: A good logging tool should provide real-time monitoring capabilities to enable you to track the flow of logs as they are generated. This can help you detect issues as they occur and take corrective action in real time.

- **Offer centralized management**: A good logging tool should provide centralized management of logs, allowing you to easily manage logs from different sources and track changes to log data over time. It should also provide access controls and permission settings to ensure that logs are accessed only by authorized personnel.

- **Support customization**: A good logging tool should be customizable to meet the specific needs of your application. This includes the ability to customize log formats and fields, as well as the ability to integrate with other tools and systems.

- **Enable correlation of logs**: A good logging tool should enable you to correlate logs from different sources and identify patterns and relationships between log data. This can help you gain deeper insights into how your application is performing and identify potential issues.

- **Provide auditing and compliance capabilities**: A good logging tool should provide auditing and compliance capabilities to help you meet regulatory requirements and internal policies. This includes features such as access controls, logging user actions, and the ability to generate audit reports.

As with monitoring tools, it is hard to recommend specific tools to use for logging, as this will vary depending on the specific tech stack in use, and also how the application will be used. Here is a list of some tools that you can research before starting to build a logging system, with the .NET/Microsoft stack tools included at the top again! Please note that Microsoft provides a logging API that is designed to work with a variety of built-in and third-party logging providers:

- **.NET built-in provider**: This is generally fine for smaller applications, but you may find that it does not provide as rich a feature set as some of the others on this list. It is a useful tool to get started with, but one that your application may quickly outgrow.

- **Serilog**: A popular structured logging library for .NET applications that supports multiple sinks and enrichers for enhanced logging capabilities.

- **NLog**: A flexible and high-performance logging library for .NET, providing advanced routing and filtering options for log events.

- **log4net**: A widely used logging library for .NET applications, inspired by the popular log4j library for Java, offering a variety of logging targets and flexible configuration options.

- **Seq**: A centralized log server and structured log data viewer, often used in conjunction with Serilog, providing powerful querying and visualization features for analyzing log events.

- **ELK Stack**: A popular open source log management platform that combines Elasticsearch for indexing and searching, Logstash for log processing and routing, and Kibana for the visualization and analysis of log data.

- **Application Insights**: A Microsoft Azure service that provides application performance monitoring, diagnostics, and logging capabilities, easily integrated into .NET applications.

- **Azure Log Analytics**: A log management and analytics service in Azure that can collect, store, and analyze log data from various sources, including application logs, Azure resources, and virtual machines.

These tools and services cater to different aspects of logging, from libraries used within the application code to centralized log management and analysis platforms. The choice of tools will depend on the specific requirements and constraints of your SaaS application, as well as your preferred development ecosystem.

How to do this

Implementing a robust logging strategy is essential for any SaaS application. While we've discussed various tools that can be used for logging, it's also important to understand the process of setting up an effective logging strategy. Here's a list of steps to follow when implementing logging in your application. Keep in mind that logging is focused on recording events that occur within your application code:

1. Identify the events and information that are important to log in your application. This may include errors, user actions, system events, and other relevant data that can help you understand the application's behavior, troubleshoot issues, and maintain an audit trail for security and compliance purposes.

2. Choose the logging tool or library that best fits your application's requirements and technology stack. There are numerous logging tools available, each with its own strengths and weaknesses. Make sure to select a tool that is compatible with your application and provides the necessary features.

3. Configure the logging tool to capture the relevant events and data identified in step 1. This may involve setting up log levels, filters, and custom log entries within your application code to ensure that you're capturing the right information.

4. Establish a consistent log format and structure across your application to make it easier to analyze and correlate log data. This may include using standardized timestamps, log levels, and message formats to ensure consistency.

5. Set up log aggregation and storage to centralize and retain log data for analysis. This may involve configuring your logging tool to send log data to a central log management system, storing logs in a database, or using a cloud-based log storage service.

6. Implement log analysis and monitoring to proactively detect issues and trends within your log data. This may involve using log analysis tools, setting up alerts based on log events or patterns, and regularly reviewing log data for insights.

7. Integrate your logging strategy with your existing development and operations workflows, such as your issue-tracking system, CI/CD pipelines, and communication platforms. This will ensure that your team is aware of any issues and can take action accordingly.

8. Continuously review and refine your logging strategy as your application evolves. As new features are added, security requirements change, or user expectations grow, you may need to adjust your logging approach to capture the necessary information.

By following these steps and tailoring your logging strategy to the specific needs of your SaaS application, you'll be well equipped to maintain a comprehensive record of your application's events, diagnose and troubleshoot issues more effectively, and ensure a secure and compliant platform for your users.

Best practices for logging

Much like monitoring, logging can be challenging to get right in a SaaS application. Here are some best practices to keep in mind when designing a logging system:

- **Define log levels**: It is important to establish clear log levels that categorize log events based on their severity or importance. These levels can include Debug, Info, Warning, Error, and Critical, and can help you identify and prioritize issues based on their impact on the application.

- **Use structured logging**: Implementing structured logging enables you to capture log events in a machine-readable format, making it easier to filter, search, and analyze log data. By including structured data in your logs, you can provide additional context and information about the event, making it easier to identify and troubleshoot issues.

- **Include context**: Ensure that your log messages provide enough context to identify the source of an issue. This can include relevant variable values, user IDs, or timestamps. By providing this information, you can more easily identify and troubleshoot issues when they occur.

- **Log correlation**: In distributed systems or microservices architectures, it can be challenging to trace the flow of requests and identify issues across multiple services. Using correlation IDs or trace IDs to link related log events can make it easier to identify and troubleshoot issues across different services and components.

- **Centralize log management**: Aggregating logs from multiple sources into a centralized log management system can provide a comprehensive view of your application's performance and make it easier to monitor and analyze log data. This can enable you to identify issues and troubleshoot problems more quickly.

- **Implement log retention policies**: Define retention policies for log data based on storage limitations, compliance requirements, and the usefulness of historical log data. By archiving or deleting logs as necessary, you can reduce storage costs and ensure compliance with regulatory requirements.

- **Secure sensitive information**: Avoid logging sensitive information, such as **Personally Identifiable Information** (**PII**) or authentication credentials, to prevent data breaches and maintain compliance with data privacy regulations. By implementing appropriate security measures, such as encryption and access controls, you can protect your log data from unauthorized access.

- **Monitor logs in real time**: Setting up real-time log monitoring and alerting can help you detect and address issues proactively. By monitoring logs in real time and setting up alerts based on specific log events, you can quickly identify and address issues before they impact users or system performance.

- **Optimize log performance**: Ensure that logging does not negatively impact the performance of the application. This can include using asynchronous logging, batching, and throttling as needed to optimize log performance and prevent log-related performance issues.

- **Review and refine**: Periodically reviewing and refining your logging strategy can help you identify areas for improvement and adjust log levels, message formats, or retention policies as necessary. By continuously improving your logging system, you can ensure that it remains effective and efficient over time.

Next, we will look at some SaaS-specific considerations that you must keep in mind.

Monitoring and logging considerations for SaaS applications

As we have discovered throughout this book, developing SaaS applications can be challenging, with numerous specific considerations when using various types of technology. In this section, we will look at monitoring and logging considerations that may be more specific to a SaaS application:

- **Multi-tenancy** is a commonly used technique when building SaaS applications. Monitoring and logging in a multi-tenant environment require careful attention to ensure the appropriate isolation of tenant data and to track tenant-specific performance metrics. Developers need to design monitoring and logging strategies that can effectively identify issues affecting specific tenants while maintaining data privacy and compliance. As we discussed in *Chapter 3*, maintaining data isolation in a multi-tenant system is both difficult and of the utmost importance. Systems that centralize data collection, such as monitoring or logging systems, can easily become a weak link in the chain if care is not taken.

- **Microservices** have become a popular architectural style for building scalable and maintainable SaaS applications. Monitoring and logging microservices require a granular approach to capture the performance, health, and events of individual services within the application. This can make building the monitoring and logging infrastructure challenging, as there could be many different services in the microservice constellation, each with its own requirements. Debugging runtime errors in a microservice application can quickly become a nightmare. Even if it adds some challenges, building robust monitoring and logging for a microservice app is extremely important.

- **Scalability** is a crucial aspect of SaaS applications, as user bases and workloads can grow rapidly. We will discuss operating at scale in detail in *Chapter 12*. Monitoring and logging systems should be designed to adapt to changes in scale, ensuring that they can continue to provide accurate and timely insights even as the application grows. This includes monitoring resource consumption, load balancing, and auto-scaling capabilities to maintain optimal performance and resource allocation. The logging system should also be capable of handling increasing data volumes and user loads.

- **Distributed architectures** involve multiple components and services working together across different physical or virtual locations. Monitoring and logging such systems require a comprehensive approach that can capture and correlate events and metrics from various sources, enabling developers to gain a holistic view of the application's health, performance, and event history. Techniques such as distributed tracing, log aggregation, and centralized monitoring can help manage the complexity of distributed architectures.

- **Integration with cloud services** is common in SaaS applications, as they often leverage services such as storage, databases, and messaging provided by cloud platforms. Monitoring and logging these integrations involve tracking the performance, availability, and usage of these cloud services, ensuring that they meet the application's requirements and SLAs. Developers should also consider the monitoring and logging capabilities and tools provided by the cloud platform itself to gain deeper insights into the integrated services.

- **Compliance** plays a critical role in SaaS applications, particularly when handling sensitive data or operating within regulated industries. Ensuring compliance means adhering to a set of established rules, standards, or regulations set forth by industry-specific organizations, government bodies, or international institutions. Monitoring and logging systems need to be designed with compliance in mind, capturing security-related metrics, events, and audit trails to demonstrate adherence to these requirements. Compliance may also dictate specific logging retention policies, access control measures, and encryption practices to protect sensitive information. By integrating compliance-focused monitoring and logging practices into your SaaS application, you not only safeguard your customers' data and privacy but also mitigate potential legal and financial risks associated with non-compliance.

- Finally, **security and compliance** are extremely important in SaaS applications, especially when handling sensitive data or operating in regulated industries. Monitoring and logging should encompass security-related metrics and events, such as authentication failures, unauthorized access attempts, and policy violations. This focus helps developers proactively identify potential security threats, maintain compliance with industry standards and regulations, and ensure that tenant-specific logging requirements or preferences are met, such as log levels, data retention policies, or alerting thresholds.

Summary

In this chapter, we have explored the critical importance of monitoring and logging in the realm of SaaS applications, particularly when considering the complexities and unique challenges that arise when working with Microsoft technologies. As SaaS applications serve millions of users, operate around the clock, and handle a diverse range of actions, implementing robust monitoring and logging systems is essential to maintaining the reliability, performance, and security of these applications.

We delved into the differences between monitoring and logging, highlighting that monitoring is a proactive technique focused on observing the system's health and performance, whereas logging is primarily concerned with recording events and data to enable effective troubleshooting and analysis. Both techniques serve as complementary tools in the arsenal of SaaS developers, ensuring a seamless and reliable user experience.

Throughout the chapter, we explored the key aspects of monitoring and logging, discussing the importance of performance metrics, resource utilization, application availability, and health, as well as the relevance of log levels, structured logging, and context. We also examined the unique considerations that arise in SaaS applications, such as multi-tenancy, microservices, scalability, distributed architectures, integration with cloud services, security, and compliance.

We covered the best practices for monitoring and logging in SaaS applications, emphasizing the significance of defining relevant metrics and thresholds, implementing proactive monitoring and alerting, ensuring data privacy and compliance in multi-tenant environments, and integrating monitoring data with logging and other diagnostic tools. Additionally, we highlighted the importance of continuously refining and improving monitoring and logging strategies to adapt to the ever-changing demands and requirements of SaaS applications.

The chapter also introduced various tools and technologies commonly used for monitoring and logging in the Microsoft development ecosystem. We discussed the utility of Application Insights, Azure Monitor, and Azure Log Analytics for monitoring purposes, and explored logging libraries such as Serilog, NLog, and log4net, as well as log management solutions such as ELK Stack.

As we conclude this chapter, it is important to remember that monitoring and logging are not static processes. To achieve success in SaaS applications, developers must continuously review, adapt, and refine their monitoring and logging strategies to respond to new challenges, changes in user behavior, and evolving technology landscapes. By doing so, they can maintain the highest levels of reliability, performance, and security for their applications, ensuring the satisfaction and trust of millions of users.

In the next chapter, we will look into building and releasing pipelines – another very important consideration when building SaaS applications!

Further reading

- What is SaaS Monitoring?: `https://www.comparitech.com/net-admin/what-is-saas-monitoring/`

- Best practices for audit logging in a SaaS business/application: `https://chrisdermody.com/best-practices-for-audit-logging-in-a-saas-business-app/`

- Logging: `https://learn.microsoft.com/en-us/dotnet/core/extensions/logging`

- log4net guide for .NET logging: `https://stackify.com/log4net-guide-dotnet-logging/`

Questions

1. What are the most important metrics to monitor for your SaaS application and why?

2. How do you balance the need for detailed logging with the need to maintain compliance with data privacy regulations?

3. What are some common challenges you've faced when implementing a logging and monitoring system and how did you overcome them?

4. What are some best practices for configuring alerts and notifications to ensure that you're alerted of issues in a timely manner without being overwhelmed by false alarms?

5. How do you ensure that your logging and monitoring systems are scalable and able to handle the increasing volume of data as your application grows?

11
Release Often, Release Early

The ability to adapt and respond to market demands quickly and efficiently is crucial for a SaaS application to be successful. With **software-as-a-service** (**SaaS**) applications, customer satisfaction and user experience are key drivers for growth and retention, and one of the most effective ways to meet these demands is by adopting **continuous integration** (**CI**) and **continuous deployment** (**CD**) practices, collectively known as CI/CD.

CI/CD is a set of development practices that emphasizes the importance of integrating code frequently, testing it continuously, and deploying updates to the application with minimal delays. By automating these processes, CI/CD helps development teams reduce the time it takes to deliver new features, improvements, and bug fixes to users while also improving the overall quality and reliability of the software.

This chapter will provide a high-level understanding of CI/CD concepts and their significance in the context of SaaS applications. The focus will be on the principles and practices that can be applied to various CI/CD tools rather than diving into detailed instructions for specific tools such as Azure Pipelines or GitHub Actions. By keeping the discussion more general, the goal is to enable readers to gain the knowledge and insights necessary to implement CI/CD processes using their preferred tools and adapt them to their unique project requirements.

Throughout the chapter, we will cover the key components of a CI/CD pipeline, including source control integration, build and release triggers, containerization, and deployment strategies. We will also discuss best practices and tips for implementing CI/CD effectively in SaaS applications, ensuring security and compliance, scaling pipelines, and monitoring and optimizing CI/CD processes.

The following are the main topics covered in this chapter:

- Understanding CI/CD
- Configuring CI/CD pipelines
- CI/CD pipelines and tools overview
- SaaS-specific considerations

By the end of this chapter, you should have a solid understanding of the importance of CI/CD in SaaS applications and be well-equipped to implement these practices in your projects, irrespective of the specific CI/CD tools you choose. This knowledge will help you create more efficient, reliable, and adaptable SaaS applications that meet the ever-changing needs of your customers.

Understanding CI/CD

CI and CD are development practices that aim to streamline the software development life cycle by automating various stages of the process. CI focuses on automating the integration of code changes, ensuring that developers merge their work frequently and that the resulting codebase is tested continuously. CD, on the other hand, automates the process of deploying the integrated and tested code into production, making new features and bug fixes available to users as quickly as possible.

In this section, we'll take a deeper look into both types of pipelines and understand how they work together to make everyone's life easier when developing a SaaS application.

CI

A CI pipeline is typically triggered when a developer submits code that has been worked on locally to a centralized repository. The purpose of the CI pipeline is to ensure that the incoming changes are compatible with the existing work and that there have been no regressions or new bugs introduced. It is typical (although not mandated) that the incoming code will not be merged into the existing code until the CI pipeline has successfully completed all of the tasks in the pipeline, which indicates that the new code is safe to merge.

The CI pipeline will typically run automatically when the new code is submitted and will carry out the following tasks:

- **Download the code**: The first thing that a CI pipeline must do is to locate and download the code. The code will typically be hosted in a Git repository somewhere in the cloud. It is common for the repo and the pipeline to sit in the same system (such as with Azure DevOps), but this does not have to be the case. For example, using GitHub to host the repo and CircleCI to run the pipelines. Before a pipeline can do anything at all, it must first get the source code!

- **Build the code**: The next thing that a CI pipeline will typically do is run a build to make sure that the newly submitted code actually compiles. This step will be performed on a **virtual machine** (**VM**) or in a container that has been configured with the tools required to build the project. This VM or container will also have to manage the dependencies, so it may require internet access so that it can download any required packages.

- **Run the tests**: Assuming that the newly submitted code builds, a CI pipeline can then move on to running the tests. As we discussed in *Chapter 9*, there are broadly three categories of tests: unit tests, integration tests, and **end-to-end** (**E2E**) tests. All of these can be run by the CI pipeline, although in some circumstances, integration and E2E tests may be skipped. Typically, at least the unit tests are always executed by the CI pipeline.

- **Test coverage reports**: If test coverage has been configured for the project, this will also be run by the CI pipeline. This can help ensure that developers are being diligent in their unit testing by ensuring that the code coverage for the application remains above a certain percentage.

- **Static analysis and linting**: If there are code standards defined by either static analysis or linting tools, these will also typically be run by the CI pipeline. This ensures that the code is written in a consistent way, regardless of who in the team has written the code.

- **Security testing**: If there are automated security tests, the CI pipeline will also run them.

- **Automatic versioning**: The CI pipeline is responsible for creating the artifacts that are subsequently released. This ensures that the versioning policy is always upheld and that every release is uniquely identifiable.

- **Artifact creation and storage**: Finally, assuming that all of the previous tasks have passed, the CI pipeline will package the built application and all of its dependencies into deployable artifacts. Typically these deployable artifacts will be used by a CD pipeline to deploy the application. The artifacts are the output from the CI pipeline and the input to the CD pipeline.

This is not an exhaustive list – there is any number of jobs that you may want to run, and ensure they pass before allowing new code to be merged into the repo. You can see that the CI pipeline can be quite busy!

CD

A CD pipeline is initiated once the CI pipeline has successfully completed its tasks and produced deployable artifacts. The purpose of the CD pipeline is to ensure that the application is deployed and released in a consistent, efficient, and automated manner, reducing the risk of human errors and minimizing the time it takes to get new features and bug fixes into the hands of users. CD pipelines typically involve several stages, such as deploying to various environments (e.g., staging, production), running post-deployment tests, and monitoring the application, as shown in the following list:

- **Deploy to environments**: The CD pipeline will usually deploy the application to different environments sequentially. It often starts with deploying to a test environment, which is a replica of the production environment. This allows the team to validate the application's behavior, performance, and compatibility with other services or components in a production-like environment.

- **Run post-deployment tests**: After the application is deployed to an environment, the CD pipeline can run additional tests, such as smoke tests or regression tests, to ensure that the deployment was successful and that the application's critical functionalities are still working as expected.

- **Monitor application performance**: The CD pipeline should include monitoring tools that gather data on the application's performance, resource usage, and error rates. This information can be used to identify potential issues or areas for improvement, helping to maintain a high-quality user experience. We talked about monitoring and logging in *Chapter 10* – some of these tasks can be initiated or at least configured by the CD pipeline.

- **Manage configuration and environment variables**: The CD pipeline should handle environment-specific configurations, such as API keys or database connection strings, ensuring that the appropriate values are used for each environment.

- **Rollback strategy**: A well-designed CD pipeline should include a rollback mechanism, allowing the team to revert to a previous version of the application if issues are detected after deployment. This can help minimize downtime and mitigate the impact of any problems. This can be particularly challenging with a database platform. Once an update has been applied to the database, it can be challenging to roll back, even with help from Entity Framework!

- **Notifications and reporting**: The CD pipeline should send notifications about the deployment status and generate reports on the deployment process. This helps to keep the team informed and enables them to quickly address any issues that arise during deployment.

The preceding tasks are just a starting point for a CD pipeline, and you can tailor the process to meet the specific needs and requirements of your application. By implementing a robust and automated CD pipeline, you can streamline the release process, improve the reliability and stability of your application, and ensure that new features and fixes are delivered to users as quickly and safely as possible.

Environments

Environments are isolated instances of an application, each with its own configurations, resources, and infrastructure. They are used to replicate various stages of the software development life cycle, allowing developers to test, validate, and optimize the application before it is released to end users. Using different environments helps to minimize risks, detect issues early, and ensure the application's stability, performance, and security.

Typically, a feature or bug fix will move sequentially through a series of environments, starting on the developer's own laptop and finally ending on the production environment, in the hands of the users. As the new code moves through the environments, the consequences of bugs increase, and so the authority required to green-light the deployment also increases. A developer on a project will hopefully have full control over the local development environment but may have no access to the production environment, which may require a senior manager's approval to make changes.

The following presents a common setup of environments in a software development process:

- **Development environment**: This environment is where developers work on their local machines, writing code and testing features as they build the application. It is configured to allow rapid iteration and debugging, and it often has relaxed security constraints compared to other environments.

- **Test environment**: The test environment is used for running various types of tests, such as unit tests, integration tests, and E2E tests. It is set up to closely resemble the production environment so that the tests can validate the application's behavior, compatibility, and performance under realistic conditions. Test environments are typically managed by **quality assurance** (**QA**) teams and are separate from developers' local machines to ensure consistent testing results.

- **Staging environment**: The staging environment is a close replica of the production environment, including configurations, infrastructure, and resources. It is used to perform final testing and validation before deploying the application to production. This environment helps to identify and address any potential issues that may not have been detected in the test environment, reducing the risk of deploying faulty software to end users.

- **Production environment**: This is the live environment where the application is made available to end users. The production environment has the most stringent security, performance, and reliability requirements, as any issues or downtime can directly impact users and the business. Deployments to the production environment should be carefully managed and monitored to ensure the application's stability and performance.

In some cases, organizations may also have additional environments, such as the following:

- **Pre-production environment**: This environment is used to perform final checks, such as load testing or security testing, before deploying to production. It is an optional environment that can be used to further reduce risks associated with releasing new software.

- **Disaster recovery environment**: This environment is a backup of the production environment, used to quickly restore the application in case of a catastrophic failure or disaster. It ensures business continuity and minimizes downtime during unforeseen events.

Using multiple environments allows organizations to detect and fix issues at various stages, improving the overall quality of the application and reducing the likelihood of deploying faulty software to end users.

Benefits of adopting CI/CD

If you think that all of this sounds like a lot of work – you are right. It is a lot of work, but there are also a number of very large benefits in building out CI/CD systems. These benefits ultimately contribute to the efficiency, reliability, and agility of your software development process, ensuring the delivery of high-quality software to your users.

CI/CD can dramatically reduce the time between the code being written and the features being in the hands of the customers. Code that has value to customers should be deployed as fast as possible, but it is very common for that code to languish for months in a repository delivering no value at all. As well as adding new features, CI/CD will result in bug fixes and patches being rolled out faster.

As well as the improved speed of deployment, CI/CD will also hugely reduce deployment failures, if not eliminate them completely. Many of the problems typically encountered during a release will be caught nearly instantly in either the CI or the CD pipeline, almost guaranteeing a successful release every time.

The benefits of CI/CD are not all for the users of the application, though. There are also significant benefits for the developers. Generally, the pipelines will enforce higher code quality and consistency through tools such as automated testing, linting, and static analysis. This makes working on the product a much more enjoyable experience for everyone involved and should ultimately result in a quicker turnaround time for new features and bug fixes on the developer side as well.

The pipelines will increase the confidence of the whole team in the releases and the ability to turn around releases quickly. This additional peace of mind is a huge benefit for the team.

We will talk about scaling the app in detail in *Chapter 12*, but it's worth mentioning at this point that CI/CD is essential when your applications start to scale, and you are required to deploy and manage multiple instances of the app to manage the load on the servers. Achieving this 'by hand' is next to impossible, and so CI/CD becomes almost mandatory when you start to seriously scale the application.

While implementing CI/CD pipelines may be a significant investment of time and resources, the benefits they provide in terms of efficiency, reliability, and overall software quality make them an essential component of modern software development processes. Embracing CI/CD will help your organization stay competitive, deliver value to users faster, and build a strong foundation for future growth.

Is CI/CD DevOps?

You will often hear CI/CD mentioned in the same sentence as DevOps, and while the concepts are related, CI/CD is not quite synonymous with DevOps. DevOps is a broader concept that encompasses the cultural shift, collaboration, and practices that bring together software development and IT operations teams with the goal of increasing efficiency, reducing the time to deliver software, and improving overall software quality.

CI/CD focuses specifically on automating the process of integrating code changes, testing, and deploying the application to production environments. By implementing CI/CD pipelines, development and operations teams can work more closely and iteratively, which aligns with the DevOps philosophy.

In short, CI/CD is a key component of the DevOps approach, but DevOps encompasses a wider range of practices, tools, and cultural shifts that aim to bridge the gap between development and operations teams.

Configuring CI/CD pipelines

While the concepts around CI/CD are fairly straightforward, they can be a deceptively large and complex topic to grasp completely. The skills that are required are something of a blend of developers and IT operations (thus: DevOps!), so it can be challenging to get right. There are a number of popular systems for doing this; the 'big three' are Azure DevOps, GitHub Actions, and Jenkins, each with its own pros and cons. In this section, I will offer some general advice that should be applicable to any system that you choose to build your pipelines in.

Source control integration

The first and most important thing to get right is to integrate the source control with the pipelines. This is foundational as without the source control, there is nothing for the pipelines to run against. This integration must allow the pipeline to detect changes to the code base (typically through a commit or a pull request) and then initiate the appropriate build, test, and deployment processes. When setting up source control integration, ensure that it supports the various branches and workflows your team uses, enabling seamless collaboration and efficient development.

Build triggers and build agents

Build triggers and build agents play a crucial role in automating the build process. Build triggers determine when the pipeline should start building the application, typically in response to events such as new code commits, pull requests, or a schedule. Build agents are responsible for executing the build tasks on dedicated machines or cloud-based environments, ensuring the application is built, tested, and packaged according to the specified configuration. When configuring build triggers and agents, consider factors such as the frequency of code changes, the resources required for building the application, and the desired level of parallelism.

A build agent is typically a VM or a container running in a cloud environment somewhere. As with everything in the cloud, you must pay per use for the build agents. This can quickly start to add up! On smaller projects, it may be possible to do a build every time code is committed to the repo, but as the team grows and the number of commits per day starts to increase, it may make more sense to do a single nightly build or even a weekly build.

Defining build tasks and stages

Defining build tasks and stages is essential for organizing and managing the various steps of the build process and keeping everything tidy. Your pipeline configuration is essentially just more code, and it should be looked after in the same way as the actual application code.

Build tasks are individual actions performed during the build, such as compiling code, running tests, or packaging the application.

Stages represent a sequence of related tasks that are executed together, often corresponding to different phases of the development life cycle, such as development, testing, and production. When defining build tasks and stages, ensure they are aligned with your team's development practices and that they support the required level of automation and testing.

Release triggers and environments

Release triggers and environments govern the deployment of the application to various environments, such as staging or production. Release triggers determine when a new release should be created and deployed, typically in response to events such as successful builds, scheduled times, or manual intervention. Environments represent the target deployment destinations, including their configurations, resources, and access controls. When configuring release triggers and environments, consider factors such as the desired release frequency, the complexity of your deployment process, and the need for staging and testing before deploying to production.

It is fairly common for deployments to test or even staging environments to happen fully autonomously or with the approval of any member of the team. Deployments to production will rarely happen fully automatically and will usually need approval from a senior member of the team. Often multiple approvals will be required, for example, from management, the QA team, and the devlopment manager.

Deploying to multiple tenants

Deploying to multiple tenants is a key aspect of SaaS applications, as it allows you to serve multiple customers using the same codebase while maintaining data isolation and customization. To achieve this, configure your CI/CD pipeline to support tenant-specific deployments, enabling you to deploy updates and new features to all tenants simultaneously or selectively. This may involve parameterizing your deployment tasks, using tenant-specific configurations, or leveraging features provided by your CI/CD tool or hosting platform.

We discussed multi-tenancy in detail in *Chapter 3*, so you will appreciate how challenging it can be to manage an application that has many tenants, some of which require fully segregated installations.

Deploying microservices in SaaS applications

Deploying microservices is a crucial aspect of SaaS applications, as they enable you to build scalable, flexible, and maintainable systems. Microservices architecture allows you to divide your application into small, independent components, each responsible for a specific function or feature. This enables you to develop, test, and deploy these components independently, reducing the complexity and risk associated with monolithic applications. We discussed microservices in detail in *Chapter 6*, so you will appreciate the benefits and challenges associated with managing a SaaS application built on this architecture.

In the context of CI/CD pipelines, deploying microservices requires careful coordination and management to ensure that each service is built, tested, and deployed in a consistent and reliable manner. This may involve configuring your CI/CD pipeline to handle multiple repositories, using service-specific build and deployment tasks, and leveraging containerization technologies, such as Docker, to package and deploy your microservices.

Additionally, deploying microservices in a SaaS application may involve integrating with other components, such as APIs, databases, and third-party services. This requires your CI/CD pipeline to manage dependencies, versioning, and configuration settings for each microservice, ensuring seamless interaction between all components of your application.

Approvals and gates for quality control

Approvals and gates for quality control are vital for ensuring that your application meets the required standards before being deployed to production. Approvals involve manual sign-offs from designated team members, while gates are automated checks that must pass before proceeding to the next stage of the pipeline. Examples of gates include successful test results, performance metrics, or security scans. By implementing approvals and gates, you can minimize the risk of deploying faulty or insecure code, ensuring that your SaaS application maintains a high level of quality and reliability.

CI/CD pipelines and tools overview

There are quite a few tools and systems available that you can use to build your CI/CD pipelines. Often, the CI/CD tool will come along with the source control tool that you are using, but that does not have to be the case, and it's worth understanding the available tools so you can make your choice.

The feature overlap of these tools is fairly extensive, and you will find that all of the main functionality that we have discussed in this chapter is available in all of the mainstream tools. The choice of tool will largely come down to individual preference. The three most commonly used tools in enterprise settings are Azure DevOps, GitHub Actions, and Jenkins. The open source community more commonly uses CircleCI or Travis CI.

Popular CI/CD tools

Microsoft's Azure DevOps is a comprehensive suite of tools that covers the entire development life cycle, from planning and coding to building, testing, and deploying. It offers a range of services, including Azure Repos for source control, Azure Boards for project management, and Azure Pipelines for CI/CD. Azure DevOps provides seamless integration with other Microsoft services and supports various programming languages, platforms, and frameworks. It is particularly suited for teams already using Microsoft technologies and looking for a tightly integrated CI/CD solution.

GitHub Actions is a CI/CD solution built directly into GitHub, making it an attractive option for teams that already use GitHub for source control. With GitHub Actions, you can create custom workflows using a variety of pre-built actions or create your own. These workflows can be triggered by various events, such as commits, pull requests, or scheduled events. GitHub Actions offers a marketplace where you can find a wide array of community-contributed actions, enabling you to quickly build and customize your CI/CD pipeline. It also supports multiple languages, platforms, and frameworks.

Jenkins is an open source CI/CD server that has been widely adopted in the software development industry. It offers a high level of flexibility and extensibility thanks to its large ecosystem of plugins and integrations. Jenkins supports various build tools, version control systems, and deployment platforms, making it a versatile option for teams with diverse technology stacks. With Jenkins, you can create custom-build pipelines using its pipeline-as-code feature, allowing you to manage your pipeline configuration within your source control system.

Travis CI is a popular CI/CD platform known for its ease of use and seamless integration with GitHub. It offers both cloud-based and on-premises options, providing flexibility to organizations with different requirements. Travis CI supports a wide range of languages, platforms, and frameworks, making it a versatile choice for various projects. Like other CI/CD tools, Travis CI enables you to define your build pipeline as code, which can be version-controlled and managed within your repository.

CircleCI is another popular CI/CD platform that emphasizes speed and simplicity. It offers a cloud-based solution as well as a self-hosted option for teams with specific security or compliance requirements. CircleCI supports a wide range of languages and platforms and provides a robust set of integrations with other development tools. Its pipeline-as-code approach, like other CI/CD tools, allows you to manage your pipeline configuration within your source control system, making it easy to maintain and update as your project evolves.

These tools, including Azure DevOps, GitHub Actions, Jenkins, Travis CI, and CircleCI, offer a diverse set of options for CI/CD. The best choice for your specific needs will depend on factors such as your existing technology stack, team size, and project requirements. Each of these tools provides unique features and benefits, so it is important to evaluate them based on your team's needs and preferences.

Factors to consider when choosing a CI/CD tool

Selecting the right CI/CD tool for your project is an essential step toward building a successful pipeline. A well-chosen CI/CD tool can improve your team's productivity, streamline your processes, and help you maintain a high-quality codebase.

One of the most critical aspects to consider when selecting a CI/CD tool is its ability to seamlessly integrate with your current source control, issue tracking, and other development tools. This ensures a smooth and efficient workflow, reducing the overhead of managing disparate systems. Before choosing a CI/CD tool, evaluate its compatibility with your existing tools and services, and consider the ease of integration.

We will discuss scaling in detail in the next chapter, but it's worth mentioning at this point that as your SaaS application grows, your CI/CD pipeline should be able to scale with it. It's essential to consider the tool's capabilities to handle large projects and multiple teams working simultaneously. A scalable CI/CD tool should be able to support increasing workloads, additional users, and more complex pipelines without compromising performance or reliability.

Depending on your project's specific requirements, you may need a CI/CD tool that offers a high degree of customization and extensibility. This could be through plugins, integrations, or custom scripting. A customizable CI/CD tool allows you to tailor the pipeline to your unique needs, implement custom logic, and integrate with niche tools or services. Consider the available options for extending the tool's functionality and the ease of implementing these customizations.

Finally, none of this comes for free! Compare the pricing models and available support options for different CI/CD tools. Factors such as the size of your team, the frequency of deployments, and your budget constraints can significantly impact your decision. Many CI/CD tools offer a range of pricing tiers, including free plans with limited features and enterprise plans with advanced capabilities and support. Additionally, consider the quality of the documentation, community support, and vendor-provided support when evaluating CI/CD tools. A tool with strong support resources can help your team troubleshoot issues and adopt best practices more effectively.

Building a flexible and adaptable CI/CD process

While selecting the right CI/CD tool is essential, it is equally important to design a CI/CD process that is flexible and adaptable to your project's unique needs. A well-structured and agile CI/CD process can improve your team's productivity, reduce the time to market for new features, and help you maintain a high-quality codebase.

Fostering collaboration and communication among team members is crucial for a successful CI/CD process. Encouraging open discussions about the pipeline and its goals helps create a shared understanding and ownership of the process. Regular meetings, code reviews, and shared documentation can facilitate better communication and collaboration, making it easier to address issues and make improvements.

Continuous improvement is an integral part of all of the development process, and CI/CD is no different. Incorporate feedback from team members and adapt to changes in project requirements or tools as needed. By regularly reviewing and refining your CI/CD processes, you can ensure they remain efficient, up-to-date, and aligned with your project's goals.

Ensure that your CI/CD pipeline is well-documented and easily understood by new team members. Clear documentation makes your pipeline more maintainable and scalable over time, reducing the learning curve for new team members and making it easier for the team to make updates and improvements. Additionally, by documenting your pipeline's configuration, best practices, and troubleshooting guides, you can create a valuable resource for your team.

Monitoring the performance and effectiveness of your CI/CD pipeline is crucial for identifying areas for improvement and optimizing the process. Use metrics such as build success rates, deployment frequency, and lead time for changes to evaluate the pipeline's efficiency and effectiveness. Regularly analyze these metrics to spot trends, detect bottlenecks, and pinpoint areas where improvements can be made. By actively monitoring and optimizing your CI/CD process, you can ensure it remains robust, efficient, and capable of meeting your project's evolving needs.

If you understand the tools that are available, make an informed choice, and build in the correct amount of flexibility, then you should be well on the way to success with CI/CD. As with everything SaaS-related, there are a few specific considerations that are worth being aware of. We'll discuss these in the next section.

SaaS-specific considerations

SaaS applications bring their own unique set of challenges and requirements. As a result, it is essential to carefully consider the specific aspects of SaaS when building and deploying these applications. This section will explore the key SaaS-specific considerations that you should keep in mind when designing and implementing your CI/CD pipelines.

Containerization

We have made use of containerization already to build the developer environment, but that is far from the only use case in the context of SaaS applications. When developing SaaS apps, containerization is particularly valuable due to the inherent complexity and scale of such systems. By packaging each microservice in a self-contained container, developers can ensure that their applications run consistently across different environments, reducing the likelihood of issues arising from discrepancies in dependencies or configurations. Furthermore, containers enable better resource utilization and make it easier to scale individual components of the application independently, resulting in more efficient and cost-effective SaaS solutions.

To leverage containerization in your SaaS application, start by creating container images for each of your app's microservices. These images are built using a Dockerfile, which defines the base image, application code, dependencies, and runtime configuration. By creating a separate image for each microservice, you can ensure that they remain isolated, allowing you to update, scale, and deploy each service independently of the others.

Managing multi-container applications can be complex, as it often involves coordinating the deployment, scaling, and communication of multiple interconnected services. To simplify this process, use orchestration tools such as Docker Compose, Kubernetes, or Amazon **Elastic Container Service** (**ECS**), which allow you to define multi-container applications using configuration files and automate the management of containerized services. These tools help you maintain a consistent application state and facilitate communication between containers, making it easier to develop and operate large-scale SaaS applications.

Integrating containerization with your CI/CD pipeline is essential for automating the build, test, and deployment processes of your containerized SaaS application. To achieve this, configure your CI/CD pipeline to build container images for each microservice whenever code changes are integrated and automatically run tests against these images to validate their functionality and performance. Once the tests pass, the pipeline should deploy the updated images to the appropriate environments using the chosen orchestration tool. By incorporating containerization into your CI/CD pipeline, you can streamline the development and deployment processes, making it easier to deliver high-quality, scalable SaaS solutions to your customers.

Upgrades

The importance of a well-planned upgrade strategy cannot be overstated for SaaS microservice applications. As these applications often serve multiple customers with diverse requirements and high expectations for uptime, a seamless upgrade strategy ensures that new features, improvements, and bug fixes can be delivered without disrupting the user experience.

Zero-downtime deployments are a critical component of a successful upgrade strategy. By minimizing the impact of updates on the availability and performance of the application, zero-downtime deployments ensure that users can continue using the service without interruptions. There are several approaches to achieving zero-downtime deployments:

- **Blue-green deployment**: This approach involves maintaining two identical production environments, referred to as "blue" and "green." At any given time, one environment is active and serving users, while the other is idle. During an upgrade, changes are deployed to the idle environment, which is then tested and verified. Once the upgrade is deemed successful, traffic is gradually switched over to the updated environment. This method allows for quick rollbacks if issues arise, as traffic can be easily redirected back to the original environment.

- **Canary releases**: In this approach, upgrades are deployed to a small subset of users, or "canary" instances, before being rolled out to the entire user base. This enables developers to monitor the performance and stability of the upgrade in a controlled manner and identify any issues before they affect all users. If the upgrade is successful, it is gradually deployed to the remaining instances.

- **Rolling updates**: Rolling updates involve deploying upgrades incrementally across instances, typically one at a time or in small groups. As each instance is updated, it is briefly taken out of service, and traffic is redirected to the remaining instances. This process continues until all instances have been upgraded. While this approach may be slower than others, it minimizes the risk of widespread issues and allows for easier troubleshooting.

Managing database schema changes in SaaS applications can be particularly challenging, as updates often need to be performed without disrupting existing data or compromising the integrity of the application. To handle these changes, consider using tools and techniques such as migrations, versioning, or feature flags, which allow for incremental and reversible updates to the database schema. Additionally, ensure that your database is designed to support multi-tenancy, allowing for seamless upgrades across all tenants.

Monitoring and rollback strategies are crucial for quickly identifying and addressing failed upgrades. By closely monitoring the performance and stability of your application during and after an upgrade, you can detect issues early and take appropriate action. Implement a rollback strategy that enables you to quickly revert to a previous stable version of the application if an issue arises during an upgrade. By having a well-defined monitoring and rollback plan, you can minimize the impact of failed upgrades on your users and maintain the high quality and reliability of your SaaS microservice application.

Security and compliance in CI/CD pipelines

In SaaS applications, ensuring the security and compliance of your software is of paramount importance, as it involves handling sensitive data and meeting industry-specific regulations. By implementing rigorous security measures and compliance checks within your CI/CD pipeline, you can safeguard your application and its users while adhering to the required standards.

To incorporate security and compliance into your CI/CD pipeline, consider the following best practices:

- **Automate security testing**: Integrate automated security testing tools, such as **static application security testing (SAST)** and **dynamic application security testing (DAST)**, into your pipeline. These tools help identify vulnerabilities and potential security risks in your code, enabling you to address them before they reach production.

- **Implement secure coding practices**: Encourage your development team to follow secure coding best practices and guidelines. This includes adhering to the principles of least privilege, input validation, and secure data storage. You can also integrate code analysis tools into your pipeline to enforce these practices and identify potential security issues.

- **Monitor and audit your pipeline**: Regularly monitor and audit your CI/CD pipeline to ensure that it remains secure and compliant. This includes checking for unauthorized access, tracking changes to your pipeline configuration, and reviewing security logs. Implementing access controls and role-based permissions can also help prevent unauthorized modifications to your pipeline.

- **Manage secrets and credentials securely**: Store sensitive data, such as API keys, passwords, and tokens, securely by using secret management tools or secure storage services. Avoid embedding these credentials in your code or configuration files, and ensure that they are encrypted both at rest and in transit.

- **Perform regular vulnerability scans and updates**: Keep your CI/CD infrastructure up-to-date by regularly scanning for vulnerabilities and applying necessary patches. This includes updating your build tools, dependencies, and runtime environments to mitigate the risk of known security issues.

- **Compliance checks**: Incorporate automated compliance checks into your pipeline to ensure that your application meets the required industry standards and regulations. This may involve validating your application against predefined compliance policies or integrating with specialized compliance tools.

By incorporating security and compliance measures into your CI/CD pipeline, you can proactively address potential risks and maintain a high level of trust in your SaaS application. This not only protects your users but also ensures that your application remains reliable and compliant with industry standards.Top of Form

Summary

In summary, this chapter covered the essential concepts and best practices for implementing CI/CD in SaaS applications. We explored the benefits of CI/CD in enhancing the development life cycle and improving the quality of SaaS apps. We discussed various CI/CD pipelines, tools, and factors to consider when choosing a CI/CD tool, emphasizing the importance of building a flexible and adaptable process.

We examined the configuration of CI/CD pipelines, including source control integration, build triggers and agents, defining build tasks and stages, release triggers and environments, deploying to multiple tenants, and incorporating approvals and gates for quality control. We also highlighted the value of containerization in SaaS applications, discussing the use of Docker and container orchestration tools to manage and deploy containerized microservices.

We delved into upgrading SaaS microservice applications, discussing the importance of a well-planned upgrade strategy and various zero-downtime deployment techniques, such as blue-green deployment, canary releases, and rolling updates. We addressed the challenges of managing database schema changes and the need for monitoring and rollback strategies for failed upgrades.

Finally, we provided best practices and tips for CI/CD in SaaS applications, emphasizing the importance of automation and testing, ensuring security and compliance, scaling CI/CD pipelines for large-scale applications, and continuously monitoring and optimizing the pipeline. By following the guidance presented in this chapter, you can build efficient and effective CI/CD pipelines that support the development and deployment of high-quality, scalable, and reliable SaaS applications.

In the next chapter, we'll look into how to scale your SaaS applications.

Further reading

- Best Practices for SaaS Businesses: `https://www.missioncloud.com/blog/five-best-practices-for-saas-businesses-deploying-devops-as-your-secret-weapon`

- CI/CD baseline architecture with Azure Pipelines: `https://learn.microsoft.com/en-us/azure/architecture/example-scenario/apps/devops-dotnet-baseline`

- How to build a CI/CD pipeline with GitHub Actions in four simple steps: `https://github.blog/2022-02-02-build-ci-cd-pipeline-github-actions-four-steps/`

- What is Jenkins? `https://phoenixnap.com/kb/what-is-jenkins`

- What is CI/CD? `https://www.synopsys.com/glossary/what-is-cicd.html`

- DevOps: `https://aws.amazon.com/devops/what-is-devops/`

Questions

1. What are the key benefits of implementing CI/CD in SaaS applications?

2. How can containerization, such as Docker, improve the development and deployment of SaaS apps?

3. Which zero-downtime deployment techniques can be used for upgrading SaaS microservice applications?

4. What factors should you consider when choosing a CI/CD tool for your SaaS application?

5. How can you ensure security and compliance in your SaaS CI/CD processes?

6. Why is monitoring and rollback planning essential for handling failed upgrades in SaaS applications?

7. What are some best practices for scaling CI/CD pipelines to accommodate large-scale SaaS apps?

12

Growing Pains – Operating at Scale

As **Software-as-a-Service** (**SaaS**) applications grow and gain more users, they inevitably face new challenges related to performance, scalability, security, and availability. These hurdles are collectively referred to as the challenges of operating at scale. So far in this book, we have delved into the fundamentals of building SaaS applications with Microsoft technologies, covering aspects such as data modeling, microservices architecture, web APIs, Entity Framework, Blazor, and secure authentication and authorization. Although we have not explicitly addressed the issue of scaling these elements, we have been laying a solid groundwork by adhering to best practices and building a robust foundation that will prove invaluable when the time comes to scale the application.

In this chapter, we will more thoroughly explore the challenges associated with operating at scale, as always, with particular emphasis on scaling SaaS applications using Microsoft technologies. We will start with a comprehensive overview of the various facets of scaling, followed by detailed insights into techniques for scaling the database, API, and **user interface** (**UI**). Additionally, we will discuss the critical importance of monitoring and alerting, the implementation of effective DevOps practices, and robust disaster recovery planning.

By examining these aspects, we aim to provide you with the necessary knowledge and tools to confidently tackle the challenges that arise as your SaaS application expands. Our goal is to equip you with a deep understanding of the intricacies of scaling, ensuring that your application remains performant, reliable, and secure even as it caters to an ever-growing user base.

The main topics covered in this chapter are the following:

- The challenges of operating at scale
- Scaling the database
- Scaling the API
- Scaling the UI

- Monitoring and alerting
- DevOps practices for scaling SaaS applications
- Disaster recovery planning

The challenges of operating at scale

It's an exciting milestone when a SaaS application needs to scale, as it means that the application is successful and is driving revenue for the business. However, with this growth comes challenges, and it's essential to be prepared for them so that the application can continue to be successful. As your user base grows, your application must remain available at all times, be able to handle the increased demand for resources, and continue to provide excellent performance and security.

The challenges of operating at scale can be broadly categorized into several areas, including infrastructure scalability, performance optimization, security and compliance, availability and uptime, cost and resource management, and planning for scaling and growth. In this section, we will explore each of these areas in detail, discussing the specific challenges you may encounter and the strategies you can use to overcome them. We will consider how these areas impact the main layers of the application, from the database to the API and, finally, the UI.

By understanding the challenges of operating at scale before you actually start to operate at scale and by developing a plan to address them, you can build and operate a successful SaaS application that can handle the demands of a growing user base.

Throughout this book, so far, we have focused on the development of the application and on running the code locally on a developer's laptop. While we have been mindful that we will have to run at scale, scale is obviously not an issue in a dev environment! Most of these tips and techniques in this chapter refer to production environments hosted on the cloud. Because we are mostly dealing with Microsoft technologies, I will focus on Azure as the cloud platform, but the general advice in this section should apply equally to other cloud providers.

Performance and response time

One of the most critical aspects of operating a SaaS application at scale is ensuring optimal performance and response time for the user. In order to deliver a performant UI, every layer under the UI must also be performant – the app is only as performant as its least performant piece! A fast and efficient experience is vital to user satisfaction, as it directly impacts their perception of your application's quality and reliability. Studies have shown that users tend to abandon slow-performing applications or websites, leading to a loss of revenue and user engagement. Consequently, maintaining high performance and fast response times is essential to retaining users and supporting the growth of your SaaS application.

As the user base grows and the data volumes grow, the demands on your application's infrastructure and resources increase proportionately. This will result in performance degradation if not properly managed. By proactively monitoring and addressing performance and response time, you can create a positive user experience that keeps customers engaged and loyal to your SaaS application. There are a few practical steps that you can take to keep on top of the application performance and, therefore, keep your users happy.

Regularly monitoring and profiling your application is essential for identifying performance bottlenecks and optimizing response times. Use performance monitoring tools, such as Application Insights for .NET applications, to collect and analyze metrics related to response times, throughput, and resource utilization. Profiling tools can help you pinpoint specific areas of the codebase that may be causing performance issues, enabling you to make targeted optimizations. All of this should be as automated as possible, ideally completely so, with an alert issued when the systems detect performance degradation.

Implement caching strategies to reduce the load on your application and database servers. Utilize various caching techniques, such as in-memory, distributed, and output caching, to store frequently requested data and serve it more quickly to users. **Content delivery networks** (**CDNs**) can also be employed to cache and serve static assets, such as images and scripts, from geographically distributed servers, thereby reducing latency and improving response times for users around the globe. This is a very complex subject and could probably take up an entire chapter on its own! As long as you, as a developer of a SaaS application, are aware of this, then you will be well placed to take advantage of it should the need arise. We will look at caching in more detail in the coming sections focusing on the database, API, and UI.

Optimize database performance by implementing proper indexing, fine-tuning queries, and using connection pooling. Regularly review and update database indexes to improve query execution times. Analyze slow-running queries and optimize them using the SQL Server Query Store or the SQL Server Management Studio's built-in performance tools.

Implement load balancing to distribute traffic evenly among multiple instances of your application, preventing any single instance from becoming a bottleneck. This can be achieved using technologies such as Azure Load Balancer or Application Gateway. As with much of this advice, load balancing must be completely automatic. Leverage autoscaling to dynamically adjust the number of application instances based on the current load. This ensures that your application remains responsive during peak times while reducing costs during periods of low usage.

Offload time-consuming tasks to asynchronous processes, which can run in the background without blocking the main application flow. This can help improve response times for user-facing operations, as they don't have to wait for these tasks to be complete. Message queues, such as Azure Service Bus or RabbitMQ, can be employed to manage and distribute these tasks among background worker services. You will remember that we looked at RabbitMQ in the chapter dedicated to microservices. This same technology, which allowed us to cleanly separate our application, can also be used to improve or maintain performance.

As you can probably tell, there are many different tricks, tools, and techniques that can be used when scaling a SaaS application!

Reliability and availability

Reliability and availability are also very important components of operating a SaaS application at scale, as they directly influence user trust and satisfaction. A reliable application consistently performs its intended functions without unexpected failures or errors, while application availability refers to the application's ability to be accessible and operational whenever users need it. Ensuring high reliability and availability is vital for user retention and building a positive reputation for your SaaS application.

As an application gains traction, it is common for growth to be non-linear, with periods of plateau followed by occasional sharp increases in demand that wither, taper out, or persist. These are challenging circumstances in which to maintain application uptime! As your application scales, it becomes increasingly important to design for fault tolerance, redundancy, and effective monitoring to minimize downtime and ensure a seamless user experience even during periods of variable demand – or sharply increasing demand!

Design your application to be fault-tolerant by implementing redundancy at various levels, including data storage, compute resources, and network connections. This can be achieved by deploying multiple instances of your application across different geographic regions or availability zones. In case of a failure in one instance, the other instances can continue to serve users without interruption.

Additionally, ensure that your data is replicated across multiple locations to safeguard against data loss and facilitate quick recovery. Services such as Azure SQL Database and Azure Storage Service provide built-in data replication features that can help you achieve this level of redundancy relatively easily.

No matter how good your systems are, there will eventually be a failure that requires you to recover some data from a backup. To make this eventuality as seamless as possible, implement regular backups of your application data and configuration to enable quick recovery in case of data loss or corruption. Use tools such as Azure Backup or SQL Server Backup to automate the data backup process and ensure that your backups are stored securely and independently from the primary data storage. Additionally, establish a disaster recovery plan to outline the steps for restoring your application in case of a catastrophic event. Don't forget to test your disaster recovery plan periodically to validate its effectiveness and make any necessary adjustments.

Doing backups alone is not enough – you should also implement a regular practice of performing recoveries, where data is restored from the backup and checked for consistency. There have been many recorded instances where the 'backup' was not as complete as it was assumed, and this fact was only discovered during a restore after a data loss.

Backups and recoveries will save you after a data loss. But measures should be taken to prevent this circumstance from occurring in the first place. Set up comprehensive health monitoring and alerting to detect and respond to potential issues before they impact users. Use monitoring tools to collect metrics, logs, and traces from your application, infrastructure, and network. Configure alerts and notifications based on predefined thresholds, allowing your team to promptly address issues and minimize downtime.

Even with the best-in-class logging and monitoring, and a solid backup and recovery strategy, there will be times (sadly) when your application buckles under the strain of a surge of new users. Design your application to degrade gracefully under heavy load or during partial failures. Implement techniques such as circuit breakers, timeouts, and retries to handle errors and failures in a controlled manner, preventing cascading failures and ensuring that users can still access core functionalities even when certain components or services are unavailable.

Security and compliance

Security and compliance are of paramount importance when operating a SaaS application at scale, as they protect your users' data, your application's integrity, and your company's reputation. A secure application safeguards sensitive data from unauthorized access, prevents malicious attacks, and maintains the confidentiality, integrity, and availability of user data. Compliance ensures that your application adheres to applicable legal, regulatory, and industry standards, mitigating risks and building trust among your users.

As your application grows, the potential attack surface increases, making it crucial to implement robust security measures and maintain compliance with relevant standards. By proactively addressing security and compliance, you can create a secure environment that protects your users and your business while scaling to meet the demands of a growing user base.

The first line of defense is to implement robust authentication and authorization mechanisms to control access to your application and its resources. We discussed this in an earlier chapter and gave some examples of how to build this into a microservice architecture with .NET. As per that example, you should never try to build your own infrastructure – always use battle-tested solutions such as OAuth 2.0, OpenID Connect, or Azure Active Directory for user authentication, and implement **role-based access control** (**RBAC**) or claims-based authorization in a standard way to enforce fine-grained permissions within your application.

There is little point in worrying about authentication and authorization if your application is transmitting data in an unencrypted way. Protect sensitive data both in transit and at rest by implementing strong encryption methods. Use encryption protocols such as **Transport Layer Security** (**TLS**) for securing data in transit and encryption technologies such as Azure Storage Service Encryption, Azure Disk Encryption, or **transparent data encryption** (**TDE**) for SQL Server to encrypt data at rest. Additionally, manage access to encryption keys securely using services such as Azure Key Vault. Take particular care when working with secrets in a developer environment. There have been many instances where the secrets for production have been accidentally leaked by committing them into a public repository!

Even though you may think that you have secured your application from the start, it is still very important to conduct regular security audits and vulnerability assessments to identify potential weaknesses in your application's security. Use tools such as Azure Security Center or third-party vulnerability scanners to detect and remediate security vulnerabilities. Additionally, perform penetration testing to simulate real-world attacks and assess your application's ability to withstand them.

Penetration testing is a complex topic that **requires** a very particular set of skills. It is often advisable to consult with domain experts to carry out **penetration (pen)** testing.

Set up continuous monitoring and logging to detect and respond to security incidents in a timely manner. Leverage tools such as Azure Monitor, Azure Sentinel, or third-party **security information and event management (SIEM)** solutions to aggregate and analyze logs from various sources, such as application, server, and network logs. Develop an incident response plan to outline the steps for identifying, containing, and recovering from security incidents, as well as communicating with affected users and stakeholders.

Finally, while the myriad of compliance requirements might seem like a needless burden, these things exist for a reason. Ensure your application adheres to relevant legal, regulatory, and industry compliance standards, such as **General Data Protection Regulation (GDPR)**, **Health Insurance Portability and Accountability Act (HIPAA)**, or **Payment Card Industry Data Security Standard (PCI DSS)**. Regularly review and update your application's privacy policy, data handling procedures, and security measures to maintain compliance. Consider using tools such as Azure Compliance Manager to track and manage your compliance requirements.

Infrastructure scalability

Infrastructure scalability used to be an enormous challenge. In the days of apps running on physical servers, the only way to scale was to call your hardware supplier and place an order for a truck full of new servers! This process could take months – it was simply not possible to react to moment-by-moment spikes in use, and outages were extremely common. The only way to cope with small spikes in demand was to have on hand extra capacity that was not used 99% of the time – an extremely expensive inefficiency for the company hosting the app!

Thankfully, in these cloud-enabled days, many of these problems are now consigned to history. There are, however, a new set of challenges that must be addressed!

Could infrastructure scalability quickly become a crucial factor when operating a SaaS application at scale, as it ensures that your application can adapt to varying demands and continue to provide a high-quality user experience? Scalable cloud infrastructure can grow or shrink dynamically to meet the changing needs of your application, allowing it to handle increasing loads without sacrificing performance, reliability, or availability. Similarly, cloud infrastructure can scale down again when demand dips, such as overnight in your most active region. This allows the operator of the application to be extremely efficient with their usage, only having to maintain a small amount of always-on buffer against spikes in use.As your application's user base and resource requirements grow, it becomes increasingly important to design and implement infrastructure that can scale both vertically and horizontally. By proactively addressing infrastructure scalability, you can create an adaptable environment that supports your application's growth and continues to meet the demands of a growing user base. Horizontal scaling refers to designing your application to run on multiple instances or nodes, which can be added or removed as needed to accommodate changing loads. To achieve this, it is very useful to embrace microservices architecture, as we have discussed in an earlier chapter. A microservice architecture allows you to scale individual components or services independently, improving resource utilization and management. It is also advisable to use containerization technologies such as Docker and orchestration platforms such as Kubernetes or **Azure Kubernetes Service (AKS)** to simplify the deployment and management of your microservices.

Vertical scaling refers to the process of increasing the resources, such as CPU, memory, or storage, allocated to your application's components as needed to handle increased demand. Regularly analyze and optimize your application's resource usage to ensure that it is using the available resources efficiently. Use tools such as Azure Monitor or Application Insights to track resource utilization and identify potential bottlenecks.

If your application is designed such that it can easily scale horizontally, and your cloud infrastructure can scale vertically, then you have given yourself the best chance of coping with transient spikes in demand!

These spikes can happen at any time, night or day, and can often happen very rapidly. There is no time to get a team on it, and both horizontal and vertical scaling must be built to happen automatically in response to the additional demand. Use automated services to define scaling rules and triggers based on predefined metrics, such as CPU utilization or request rate. Combine autoscaling with load balancing, using technologies such as Azure Load Balancer or Application Gateway, to distribute traffic evenly across instances and ensure optimal performance and resource utilization.

One very modern and also very clever approach to help facilitate automatic scaling is to adopt **Infrastructure-as-Code (IaC)** practices to automate the provisioning, configuration, and management of your infrastructure. IaC allows you to define your infrastructure as code, version control it, and consistently apply changes across environments. Use tools such as **Azure Resource Manager (ARM)** templates, Terraform, or Ansible to implement IaC and streamline your infrastructure management.

Finally, again, no matter how good your processes and practices are, there will inevitably be some unexpected issues that arise. To mitigate the impact of this, continuously monitor your infrastructure's performance, resource utilization, and capacity to make informed decisions about scaling. Use monitoring tools such as Azure Monitor, Application Insights, or third-party solutions to collect and analyze infrastructure metrics. Regularly review capacity planning to estimate future resource needs and ensure that your infrastructure is prepared to handle expected growth. By doing so, you give yourself the best possible chance to catch issues before they occur, or at least respond very quickly when they do!

Cost and resource management

In the previous section, we talked about horizontal and vertical scaling by adding additional resources for your application to consume. Even when talking about cloud infrastructure, adding resources costs additional money and can become extremely expensive as your application scales.

Therefore, efficient cost and resource management is essential when operating a SaaS application at scale, as it enables your organization to optimize the use of resources, reduce expenses, and maintain a sustainable and profitable business model. As your application's user base and infrastructure grow, it becomes increasingly important to implement strategies that help you monitor, control, and optimize the costs associated with running and scaling your application.

By proactively addressing cost and resource management, you can create an adaptable and cost-effective environment that supports your application's growth while maximizing return on investment.

This starts by simply paying attention to the costs as they arise. Regularly analyze and optimize your application's resource usage to ensure that it's using the available resources efficiently. Use monitoring tools such as Azure Monitor, Application Insights, or third-party solutions to track resource utilization and identify potential bottlenecks or underutilized resources. Implement autoscaling and load balancing strategies, as discussed in the *Infrastructure scalability* section, to optimize resource allocation and utilization.

As with much of the advice in this chapter, it is important to continuously monitor your application's costs using tools such as Azure Cost Management, AWS Cost Explorer, or third-party cost management solutions. Set up cost alerts and notifications to keep your team informed about cost trends and potential budget overruns. Regularly review and analyze cost reports to identify opportunities for cost optimization and to ensure that your application's expenses are in line with your budget and business goals.

Choosing the right infrastructure and resources for your application based on its specific requirements and usage patterns is challenging and is often overlooked by technical teams that just want to build cool applications! But, the success of an application is ultimately a function of its profitability, so care should be taken to choose the most appropriate cloud services. Regularly review your infrastructure choices and right-size your resources to ensure that you're not overprovisioning or underutilizing resources.

Data consistency and integrity

Data consistency and integrity are critical aspects of operating a SaaS application at scale, as they directly impact the quality and reliability of the data your application processes and stores. Ensuring data consistency means that your application presents a coherent view of the data to all users, regardless of where the data is stored or accessed. Data integrity refers to maintaining the accuracy, completeness, and consistency of the data over its entire life cycle.

As your application's user base and data volume grow, it becomes increasingly important to implement strategies that ensure data consistency and integrity across your application's components and services. By proactively addressing data consistency and integrity, you can create a dependable environment that maintains the quality of your data and supports your application's growth.

When building a SaaS app, it is very common to be working with distributed data systems or microservices. With these technologies, you should consider adopting an eventual consistency model to maintain data consistency across multiple data stores or services. In this model, data updates are allowed to propagate asynchronously between different components, eventually reaching a consistent state. Implement mechanisms such as message queues (such as RabbitMQ, as demonstrated in the microservices chapter) or event-driven architectures to propagate data updates and enforce consistency across your application's services.

Having a solid data model at the database layer is extremely important, but it is also important to try to prevent bad data from entering the database in the first place. To achieve this, implement data validation and sanitization processes at the UI, API, and database levels to ensure that only accurate and well-formed data is stored and processed. Use input validation techniques, such as data type constraints, range checks, and pattern matching, to validate incoming data before it's stored or processed. Additionally, sanitize data to remove any potentially harmful content or formatting, thereby preventing security vulnerabilities such as SQL injection or **cross-site scripting (XSS)** attacks.

As discussed in the *Reliability and availability* section, regularly back up your application data to protect against data loss or corruption. Implement backup strategies that include multiple levels of redundancy, such as full, differential, and incremental backups. And don't forget to test your backup and recovery processes periodically to ensure they are effective and can restore data integrity in case of a failure.

A common theme across all of these considerations for scaling is to continuously monitor and audit your application's data operations to detect and respond to potential issues that may impact data consistency and integrity. Regularly review data audit logs to identify trends and patterns, as well as to ensure compliance with relevant regulations and standards.

Planning for scaling and growth

So far, this chapter has focused mostly on technical tips, but it is also important to consider the non-technical elements that are involved in scaling the application so that it can handle changes in demand. Putting plans in place to scale and grow is a vital aspect of operating a successful SaaS application at scale, as it ensures that your application is prepared to handle the demands of an expanding user base and can continue to deliver a high-quality user experience. By proactively planning for growth, you can create an agile and resilient environment that supports your application's growth and helps maintain a high level of customer satisfaction.

The first step is to just sit down with the technical team and any other stakeholders to regularly review your application's capacity planning and resource requirements to estimate future needs based on historical trends, user growth projections, and resource utilization patterns. Nothing stays static for long in tech, so update your capacity plan periodically to ensure that your application and infrastructure are prepared to handle expected growth.

To validate your assumptions and to provide input into the growth-planning sessions, conduct regular performance testing and benchmarking to assess your application's ability to handle increasing workloads and user concurrency. Use load testing and stress testing tools to simulate real-world usage scenarios and identify potential bottlenecks or performance issues. Establish performance baselines and set target metrics to help guide your scaling efforts and ensure that your application continues to meet performance requirements as it scales.

Of course, problems will occur along the way. The less blindsided the team is by these issues, the less of an impact they will have, so develop a comprehensive disaster recovery and business continuity plan to ensure that your application can recover from unexpected failures and continue to provide service to your users. As discussed in the *Reliability and availability* section, implement backup and recovery strategies, redundancy, and failover mechanisms to minimize downtime and data loss. Regularly test and update your disaster recovery plan to ensure that it remains effective and aligned with your application's growth and evolving requirements.

It is very easy to focus on the technical challenges that are involved in scaling an application, but that should not be the only consideration. Planning for the future in this respect will prove invaluable as your application grows!

Embracing DevOps and automation

Everything that we have discussed in this chapter so far is predicated on a solid understanding of DevOps and automation. It is virtually impossible to keep up with the ever-changing demands placed on a modern SaaS application when using manual processes.

Embracing DevOps and automation enables your team to streamline development and operations processes, increase efficiency, and minimize potential risks. By integrating development and operations teams and leveraging automation tools and practices, you can ensure that your application remains agile, reliable, and adaptable as it grows.

By proactively incorporating DevOps and automation into your organization's culture and processes, you can create a collaborative and efficient environment that supports your application's growth and helps maintain a high level of customer satisfaction.

The core of this is **continuous integration and continuous deployment** (**CI/CD**), which we will discuss in detail in the coming chapter! CI/CD pipelines to automate the process of building, testing, and deploying your application are foundational to the process as they significantly reduce the time and effort required to release new features and improvements while minimizing the risk of introducing errors, regressions, or performance issues.

The CI pipelines should always verify the correctness of the code by running a suite of automated tests. This includes unit, integration, and end-to-end tests. Automated testing paired with robust CI pipelines massively reduces the risk of introducing errors or performance issues as your application scales.

Modern cloud infrastructure allows us to adopt IaC practices to manage and provision your application's infrastructure using code and configuration files rather than manual processes. IaC enables you to automate infrastructure provisioning and configuration, ensuring consistency, repeatability, and scalability. Tools such as Terraform can be used to facilitate this.

As we have demonstrated with the demo application environment in this book, it is also possible to configure **Developer Environment as Code** (**DEaC**) and build all of the developer dependencies into a Docker setup.

Continuing the theme of 'automate everything', is also very handy to implement configuration management tools and practices to automate the process of managing your application's settings, dependencies, and environment configurations. Configuration management helps ensure consistency and reliability across your application's components and services while simplifying the process of deploying updates and scaling your infrastructure. Automating configuration also minimizes the risk that important config details for the production environments will be accidentally shared or pushed to a less secure environment.

Finally, there is a huge non-technical component to getting DevOps right. Foster a collaborative culture between development and operations teams by encouraging open communication, shared goals, and joint problem-solving. Implement tools and practices that facilitate collaboration and information sharing, such as project management tools such as Jira or Trello and communication platforms such as Microsoft Teams or Slack. Regularly hold cross-functional meetings and retrospectives to review progress, discuss challenges, and identify opportunities for improvement.

DevOps has exploded in recent years and for good reason. DevOps practices play a crucial role in the successful operation of SaaS applications at scale. By bringing together development and operations teams, DevOps facilitates seamless collaboration and ensures that software is delivered quickly, reliably, and securely. With DevOps, developers can continuously deploy new features and updates, while operations teams can maintain the high availability and reliability of the application. This is especially important when operating at scale, where any downtime or interruption can have a significant impact on the user experience and revenue. The use of DevOps practices is, therefore, essential for ensuring the smooth operation of SaaS applications at scale.

In conclusion, operating a SaaS application at scale presents numerous challenges that development teams must address to ensure the continued success and growth of the application. By understanding and proactively addressing these challenges, you can create an environment that is scalable, efficient, and resilient, allowing your application to thrive as its user base expands.

Throughout this section, we have explored key areas, including performance and response time, reliability and availability, data consistency and integrity, security and compliance, infrastructure scalability, cost and resource management, planning for scaling and growth, and embracing DevOps and automation. By implementing the practical tips and strategies provided in this section, your team can tackle the challenges of operating at scale, maintain a high level of customer satisfaction, and drive the ongoing success of your SaaS application.

As you continue to grow and scale your SaaS application, it is important to regularly review and adjust your strategies and practices in response to changing requirements, new technologies, and evolving user expectations. By staying agile, adaptable, and focused on continuous improvement, your development team can successfully navigate the challenges of operating at scale and ensure the long-term success and sustainability of your SaaS application.

We will now look at particular scaling considerations across the various layers of the application.

Scaling the database

In this section, we will delve into the critical task of scaling the database layer of your SaaS application. As the foundation upon which your application is built, the database plays a pivotal role in the overall performance, reliability, and scalability of your system. Effectively managing the database becomes increasingly important as your application experiences growth, handling larger data volumes and more user requests. We'll discuss essential strategies and techniques, including sharding, horizontal scaling, caching, partitioning, archiving, indexing and query optimization, connection pooling, and replication. By mastering these approaches, you'll strengthen the database foundation and ensure a performant, scalable, and resilient SaaS application that meets the demands of a growing user base.

Sharding

Sharding is a database scaling technique that involves dividing a large dataset into smaller, more manageable pieces called shards. Each shard contains a portion of the data and is stored on a separate database server, thereby distributing the load and improving overall performance. Sharding can be particularly beneficial for SaaS applications, where the ability to handle increasing data volumes and user demands is crucial for growth and success.

There are two primary approaches to sharding:

- **Horizontal sharding (data partitioning)**: This approach divides the dataset by rows, with each shard containing a distinct subset of records. Horizontal sharding is typically based on a specific attribute, such as user ID or geographic location.

- **Vertical sharding (schema partitioning)**: In this approach, the dataset is divided into columns, with each shard containing a subset of the table's attributes. Vertical sharding is often used when certain columns are accessed more frequently or have different scaling requirements than others.

When implementing sharding, it's essential to select an appropriate sharding key to determine how data will be distributed across shards. The choice of the sharding key can significantly impact performance, so it's important to consider factors such as query patterns, data distribution, and scalability requirements. Common sharding strategies include the following:

- **Range-based sharding**: Data is partitioned based on a range of values for the sharding key (e.g., date ranges or alphabetical ranges).

- **Hash-based sharding**: A hash function is applied to the sharding key, and data is distributed across shards based on the resulting hash value. This approach typically provides a more uniform distribution of data.

- **Directory-based sharding**: A separate lookup service or directory is used to map sharding keys to specific shards, providing greater flexibility in data distribution and shard management.

While sharding can significantly improve database performance and scalability, it's important to be aware of potential challenges and considerations:

- **Data consistency**: Ensuring consistency across shards can be complex, particularly in distributed transactions or when dealing with eventual consistency models.

- **Query complexity**: Sharding can increase query complexity, as some queries may need to be executed across multiple shards and their results combined.

- **Rebalancing and resharding**: As your application grows, you may need to redistribute data across shards or add new shards. This process, known as rebalancing or resharding, can be time-consuming and may require careful planning and execution.

- **Cross-shard operations**: Operations that span multiple shards, such as joins or transactions, can be more complex and less performant than those within a single shard.

Scaling

In contrast to sharding, which involves partitioning the data into smaller subsets and distributing them across separate database servers, scaling focuses on increasing the capacity of the database infrastructure to handle increased workloads. There are two primary methods for scaling databases: horizontal scaling and vertical scaling.

Horizontal scaling, also known as scaling out, involves adding more servers or nodes to your infrastructure to handle the increased load and improve performance. In the context of databases, horizontal scaling involves replicating the entire database across multiple servers or nodes and distributing the load among them. Load balancing and data replication techniques are often employed to achieve horizontal scaling.

Vertical scaling, or scaling up, involves increasing the capacity of an existing server by adding more resources, such as CPU, memory, and storage, to handle the increased workload and improve performance. When vertically scaling a database, you upgrade the hardware or increase the resources allocated to the database server. This can involve upgrading to a more powerful server, adding more **random-access memory** (**RAM**), increasing storage capacity, or allocating more CPU cores.

Both horizontal and vertical scaling have their advantages and limitations. Horizontal scaling allows for better fault tolerance and potentially greater overall capacity, while vertical scaling can provide immediate performance improvements without the complexity of managing multiple servers. However, vertical scaling has inherent limitations in terms of resource availability and potential single points of failure.

These scaling techniques are important techniques for increasing the capacity of your database infrastructure to handle growing workloads. By understanding the differences between these methods and their respective advantages and limitations, you can make informed decisions about the best approach for scaling your SaaS application's database layer.

Partitioning

Earlier, we discussed sharding as a technique for distributing data across multiple database systems or clusters to achieve greater scalability and fault tolerance. Partitioning, on the other hand, is a related but distinct concept that involves dividing a large table within a single database system into smaller, more manageable pieces based on specific criteria. While both partitioning and sharding aim to improve performance and manageability, partitioning operates within a single database system and is transparent to the application, whereas sharding requires explicit management and coordination across multiple database systems.

Partitioning is a technique for managing large datasets in a database by splitting the data into smaller, more manageable pieces. This approach can help improve the performance, maintainability, and scalability of your SaaS application's database layer. Partitioning can be applied at both the table and index level, depending on the specific database system being used.

There are two primary types of partitioning to consider when scaling your database:

- **Horizontal partitioning**: As discussed earlier in the sharding section, horizontal partitioning involves splitting a table's rows into smaller subsets based on specific criteria, such as a range of values or a hashing function. Each partition contains a distinct subset of rows and can be stored on separate database servers or tablespaces, which can improve performance by allowing parallel processing and reducing contention.

- **Vertical partitioning**: In vertical partitioning, the table's columns are split into smaller subsets, with each partition containing a subset of the columns. This approach can be particularly useful for large tables with many columns or when specific columns are frequently accessed together. Vertical partitioning can help reduce the amount of **input/output (I/O)** required to fetch data, thus improving query performance.

When implementing partitioning, several factors should be taken into account:

- **Partitioning key**: Choose an appropriate partitioning key that provides a balanced distribution of data across the partitions. A poorly chosen key may lead to skewed data distribution and negatively impact performance.

- **Partitioning scheme**: Determine the most suitable partitioning scheme for your data based on factors such as data access patterns, query performance requirements, and maintenance considerations.

- **Data management**: Implement data management strategies, such as partition maintenance, to ensure that your partitions remain optimized and up to date. This may involve tasks such as adding or merging partitions, reorganizing partitions, or updating partition statistics.

- **Query optimization**: Optimize your queries to take advantage of partitioning, using features such as partition elimination and partition-wise joins, which can significantly improve query performance.

Partitioning is an effective technique for managing large datasets and improving the performance and scalability of your SaaS application's database layer. By understanding the different types of partitioning and their associated considerations, you can implement partitioning strategies that optimize query performance, facilitate data management, and enable your database to scale as your application grows.

Caching

Caching is a technique used to improve the performance and responsiveness of a SaaS application by storing frequently used data or the results of resource-intensive operations in a temporary storage area known as a cache. By using a cache, the application can quickly retrieve the data without having to recompute or re-fetch it from the database, thus reducing the load on the database and minimizing response times.

There are several caching strategies that you can employ to optimize your SaaS application's database performance:

- **In-memory caching**: This approach involves storing frequently accessed data in the application server's memory, allowing for faster data retrieval. In-memory caching can be implemented using built-in .NET caching mechanisms or third-party libraries such as Redis.

- **Distributed caching**: In a distributed caching setup, the cache is stored across multiple servers, often using a dedicated caching service such as Redis or Memcached. This approach is particularly useful for large-scale applications, as it allows the cache to scale horizontally and maintain consistency across multiple application servers.

- **Database caching**: Database caching involves using built-in caching mechanisms provided by the database system itself, such as SQL Server's buffer cache or Azure SQL Database's in-memory **online transaction processing** (**OLTP**) feature. This approach helps optimize query performance by reducing the time needed to fetch data from disk.

- **Query result caching**: By caching the results of frequently executed queries, you can reduce the need for repeated database queries and improve performance. This can be achieved using application-level caching or by leveraging database-level caching features, such as SQL Server's Query Store.

When implementing caching, it's essential to consider the following factors:

- **Cache invalidation**: Determine when and how cached data should be invalidated or updated to ensure the application serves accurate and up-to-date information

- **Cache expiration**: Define appropriate expiration policies for cached data to prevent stale data from being served to users and to optimize cache usage

- **Cache granularity**: Choose the appropriate level of granularity for caching, balancing the need for performance improvements against the potential complexity of managing fine-grained cache entries

- **Monitoring and metrics**: Implement monitoring and metrics to track cache performance, hit rates, and resource usage, allowing you to optimize caching strategies and make informed decisions about capacity planning and scaling

Caching is a powerful technique for improving the performance and scalability of your SaaS application's database layer. By understanding the various caching strategies and their associated considerations, you can effectively reduce database load, minimize response times, and provide a better overall experience for your users.

Indexing and query optimization

We touched on this in the database chapter earlier in this book. Indexing and query optimization are essential aspects of scaling your SaaS application's database, as they help ensure that your database queries run efficiently and minimize the impact on performance. Inefficient queries can have an enormous impact on the performance of an application and can also significantly increase the cost of the cloud resources running the database. It is doubly important to get this right!

An index is a database object that helps speed up the retrieval of rows from a table by providing a more efficient access path to the data. Indexes can be created on one or more columns of a table, and they enable the database engine to quickly locate the required rows without having to perform a full table scan. Creating the right indexes for your application can significantly improve query performance and reduce the load on the database.

Here are the types of indexes:

- **Clustered index**: A clustered index determines the physical order of data storage in a table. There can be only one clustered index per table, and it can significantly improve the performance of queries that retrieve data in the order defined by the index

- **Non-clustered index**: A non-clustered index stores a separate copy of the indexed columns, along with a reference to the corresponding rows in the table. You can create multiple non-clustered indexes per table, and they can help improve the performance of queries that filter, sort, or join data based on the indexed columns

- **Columnstore index**: A columnstore index stores data in a columnar format, which can provide significant performance improvements for analytical queries and large-scale data aggregation tasks. Columnstore indexes are particularly well-suited for data warehousing and reporting scenarios

As well as indexing, optimizing your queries is an important aspect of database performance tuning, as it ensures that your application retrieves data from the database efficiently. Some techniques for query optimization include the following:

- **Use the appropriate query types**: Choose the most efficient query type for your specific use case, such as using the `SELECT` statements with specific columns rather than `SELECT *`

- **Utilize indexes**: Ensure that your queries take advantage of existing indexes and consider creating additional indexes to support frequently executed queries

- **Limit the result set**: Use the `LIMIT`, `OFFSET`, or `TOP` clauses to restrict the number of rows returned by a query, which can help reduce the amount of data transferred and processed by the application

- **Optimize JOIN operations**: Minimize the number of JOIN operations in your queries and use the appropriate join types, such as `INNER JOIN` or `OUTER JOIN`, based on your data requirements

- **Analyze query plans**: Use tools such as SQL Server's Query Analyzer or Azure SQL Database's Query Performance Insight to analyze query execution plans and identify potential bottlenecks or inefficiencies

Indexing and query optimization play a crucial role in improving the performance and scalability of your SaaS application's database layer. By understanding the different types of indexes and employing effective query optimization techniques, you can ensure that your application retrieves data efficiently, minimizing the impact on database performance and providing a better user experience.

Data archiving and retention

As your SaaS application grows, the volume of data stored in your database will inevitably increase, potentially leading to slower performance and higher storage costs. Implementing a data archiving and retention strategy can help you manage the growth of your data while ensuring that your application remains responsive and cost-effective.

Data archiving involves moving historical or infrequently accessed data from your primary database to a separate, more cost-effective storage system. This process allows you to maintain the performance of your primary database by reducing the amount of data it needs to manage and query. Archived data can still be accessed when needed, albeit at a potentially slower rate, and can be used for reporting, analytics, or compliance purposes.

When implementing a data archiving strategy, consider the following factors:

- **Identify the data to be archived**: Determine which data can be safely moved to an archive without impacting your application's functionality or user experience. This may include historical transaction data, completed projects, or inactive user accounts.

- **Choose the appropriate storage solution**: Select a storage solution that meets your cost, performance, and compliance requirements, such as Azure Blob Storage, Azure Data Lake, or other archival storage services.

- **Automate the archiving process**: Implement a process to periodically move eligible data from your primary database to the archival storage system, ensuring that your data remains up-to-date and your primary database stays lean.

Data retention is the practice of defining how long data should be stored in your database or archival storage system before it is permanently deleted. A well-defined data retention policy helps you manage storage costs, comply with data protection regulations, and reduce the risk of data breaches.

When developing a data retention policy, consider the following factors:

- **Understand your legal and regulatory obligations**: Determine the minimum and maximum retention periods for different types of data based on your industry, jurisdiction, and any applicable regulations, such as GDPR or HIPAA.

- **Define retention periods based on business requirements**: Establish retention periods for each type of data based on your business needs, taking into account factors such as data value, access frequency, and storage costs.

- **Implement data deletion processes**: Develop processes to automatically delete data that has reached the end of its retention period, ensuring that your data storage remains compliant with your retention policy.

A well-executed data archiving and retention strategy can help you manage the growth of your SaaS application's data while maintaining database performance and controlling storage costs. By carefully considering which data to archive, selecting the appropriate storage solutions, and implementing a clear data retention policy, you can ensure that your application remains scalable and cost-effective as it grows.

Scaling the database is a critical aspect of ensuring the success and growth of your SaaS application. As your user base grows and the volume of data increases, it's essential to implement strategies that will help you maintain performance, reliability, and cost-effectiveness.

In this section, we covered various techniques and best practices for scaling your database, including sharding, horizontal and vertical scaling, caching, partitioning, data archiving and retention, monitoring, and performance tuning. Each of these approaches has its own advantages and trade-offs, and the specific combination of techniques that work best for your application will depend on your unique requirements and constraints.

As you continue to build and scale your SaaS application, keep these strategies in mind and continue to evaluate and adjust your approach as needed. By proactively addressing the challenges of scaling your database and adopting the right mix of techniques, you can ensure that your application remains performant, reliable, and cost-effective, providing a high-quality experience for your growing user base.

Scaling the API

In this section, we will explore the specific considerations for scaling the API in your SaaS application. A well-designed API is crucial for maintaining the performance, reliability, and flexibility of your application as it grows. As you have already built the Good Habits demo application and have implemented a microservice architecture with WebAPI, Ocelot as the API gateway, and RabbitMQ for asynchronous communication, you have laid a strong foundation for scaling your API! However, there are additional aspects you need to consider to ensure that your API remains responsive and efficient as the demands on your system increase. We will discuss various strategies and best practices, such as load balancing, API versioning, rate limiting, caching, and monitoring. By understanding and implementing these techniques, you can effectively scale your API to meet the needs of your growing user base and continue to deliver a high-quality experience for your customers.

Load balancing and API gateway optimization

Load balancing is a crucial aspect of scaling your API, as it helps distribute incoming requests evenly across your available resources, ensuring that no single instance becomes a bottleneck. By implementing load balancing and optimizing your API gateway, you can improve the performance and reliability of your API as your application scales.

Here are a few load-balancing strategies that you may want to consider:

- **Round robin**: This distributes requests evenly across all instances of your API, regardless of their current load or response time. This approach is simple and easy to implement but may not account for differences in instance performance or capacity.

- **Least connections**: This routes requests to the instance with the fewest active connections. This strategy helps ensure that instances with fewer connections can handle more requests, potentially improving overall performance.

- **Latency-based**: This directs requests to the instance with the lowest latency or response time. This approach can help minimize the impact of network latency on your API's performance.

There is a lot involved in API gateway optimization, and it's outside the scope of this book to go into detail. Here are some general points to consider:

- **Connection pooling**: By reusing existing connections between the API gateway and your API instances, you can reduce the overhead of establishing new connections, resulting in improved performance and reduced latency

- **Caching**: Implement caching at the API gateway level to store and serve frequently accessed data or responses, reducing the load on your API instances and improving response times

- **Rate limiting**: Enforce rate limits at the API gateway level to protect your API instances from being overwhelmed by excessive requests from individual clients or malicious attacks

- **Security**: Implement security features such as authentication, authorization, and API key management at the gateway level, offloading these responsibilities from your API instances and improving their performance

By employing load-balancing strategies and optimizing your API gateway, you can efficiently distribute incoming requests, improve the performance and reliability of your API, and ensure a high-quality experience for your growing user base.

API versioning and backward compatibility

As your SaaS application evolves and new features are added, changes to the API may be necessary. Ensuring backward compatibility and managing API versioning are essential aspects of scaling your API to maintain a consistent and reliable experience for clients and users.

We have covered API versioning strategies in the API chapter. Here's a quick reminder of some of the key strategies:

- **Uniform Resource Identifier (URI) versioning**: Include the API version in the URI, such as `/v1/users` or `/v2/users`. This approach is simple and easy for clients to understand but may lead to cluttered URIs and requires careful management of resources and routing

- **Query parameter versioning**: Include the API version as a query parameter, for example, `/users?version=1` or `/users?version=2`. This method keeps the URI clean and allows for more flexible versioning but it may be less intuitive for clients

- **Header versioning**: Use custom headers to specify the API version, such as `X-API-Version: 1` or `X-API-Version: 2`. This approach keeps the URI clean and separates versioning concerns from resource representation but may be less discoverable for clients

Once an API is in production, it is very important not to introduce any breaking changes that may cause errors in any consuming application. To make sure that the API remains backward compatible, you could consider the following:

- **Avoid breaking changes**: Whenever possible, design your API changes to be backward compatible, allowing existing clients to continue functioning without modifications

- **Deprecation strategy**: If breaking changes are necessary, provide a clear deprecation strategy and timeline to inform clients when older API versions will no longer be supported

- **Graceful degradation**: Implement fallback mechanisms for new API features, allowing clients that do not support the latest version to continue functioning with reduced functionality or features

- **Documentation**: Maintain clear and comprehensive documentation for each API version, helping clients understand the differences between versions and the migration process

By managing API versioning and ensuring backward compatibility, you can minimize disruptions for your clients and users while continuing to evolve and scale your SaaS application. This approach allows you to maintain a consistent and reliable experience as your API grows and adapts to changing requirements.

Rate limiting and throttling

As your SaaS application scales and attracts more users, the number of requests to your API will also grow. Implementing rate limiting and throttling strategies helps prevent abuse, protect your API from excessive load, and ensure fair usage among clients.

If your application is coming under intermittent heavy load, you could consider the following rate-limiting strategies:

- **Global rate limiting**: This sets the maximum number of requests allowed across all clients within a specified time period. This approach can help protect your API from excessive load but may not account for individual client usage patterns.

- **Per-client rate limiting**: Set a maximum number of requests allowed for each client within a specified time period. This strategy can help ensure fair usage among clients but may require more sophisticated tracking and enforcement mechanisms.

- **Tiered rate limiting**: Offer different rate limits based on client subscription levels or access tiers. This method allows you to provide differentiated service levels and encourages clients to upgrade to higher tiers for better API access.

In addition to the preceding rate-limiting strategies, you could consider the following throttling techniques:

- **Leaky bucket**: Implement an algorithm that accumulates incoming requests and processes them at a fixed rate. This approach smooths out request spikes and ensures that your API does not become overwhelmed.

- **Token bucket**: Use tokens to regulate the rate at which clients can make requests. Clients must have a token to make a request, and tokens are generated at a fixed rate. This method allows for more flexibility and adaptability in handling request bursts.

- **Exponential backoff**: Encourage clients to progressively increase the time between retries when they encounter rate limits or errors. This technique helps distribute retries over time, reducing the chance of overwhelming your API.

By implementing rate limiting and throttling strategies, you can protect your API from excessive load, prevent abuse, and ensure a high-quality experience for your users. These techniques help maintain the performance and reliability of your API as your SaaS application grows and serves a larger user base.

Caching strategies for API performance

We have discussed caching at the database layer, and we will now cover caching at the API layer. Caching is an essential technique to improve the performance and responsiveness of your API, especially as your SaaS application scales. By storing and serving frequently accessed data or responses from the cache, you can reduce the load on your API instances and improve response times. Caching at the API layer means that there is no contact with the database layer at all, so the benefits are felt throughout the stack.

The following are examples of caching strategies:

- **Client-side caching**: This encourages clients to cache API responses locally by providing appropriate cache-control headers (e.g., **Cache-Control**, **ETag**). This approach reduces the number of requests to your API and offloads caching responsibility to the clients.

- **Server-side caching**: This stores frequently accessed data or responses on the server-side, either in-memory or using an external caching service (e.g., Redis or Memcached). This method can significantly improve the performance of your API by reducing the need for time-consuming data retrieval or processing.

- **Edge caching**: This utilizes CDNs to cache and serve API responses closer to the clients. This approach can help reduce latency and improve response times, especially for clients located far from your API instances.

- **Cache invalidation**: This implements strategies to invalidate cache entries when the underlying data changes, ensuring that clients receive up-to-date information. Techniques such as cache expiry, cache versioning, or event-driven cache invalidation can be employed to maintain data consistency.

By incorporating caching strategies into your API, you can improve performance, reduce latency, and minimize the load on your backend systems. As your SaaS application scales and serves more users, effective caching becomes increasingly important to ensure a high-quality experience for your clients and users.

Asynchronous communication and message queuing

SaaS applications are often complex and require computationally expensive API calls. These can negatively impact its performance and responsiveness, as well as sharply increase the cost of cloud resources. Implementing asynchronous processing and background jobs can help offload these tasks from the main API request/response cycle, ensuring a smooth experience for your users.

In order to keep your application running smoothly, you could consider these asynchronous processing strategies and techniques for running jobs in the background:

- **Message queues**: This utilizes message queues (e.g., RabbitMQ, Azure Service Bus) to decouple the API from processing tasks. Clients send requests to the API, which then pushes tasks onto a queue for processing by dedicated worker services.

- **Event-driven architecture**: This implements an event-driven architecture to trigger processing based on specific events or actions within your system. This approach enables you to build scalable and resilient systems that can evolve with your application's needs.

- **Scheduled jobs**: This schedules background jobs to run at specific intervals, such as nightly data processing, weekly report generation, or daily cleanup tasks. This technique helps you distribute the load on your system more evenly over time.

- **Priority queuing**: This assigns different priority levels to tasks in your background job queues, ensuring that critical tasks are processed first. This approach helps you manage system resources more effectively and improve the overall user experience.

- **Retry and fallback mechanisms**: This implements retry and fallback mechanisms for background jobs that may fail due to transient errors, such as network issues or temporary resource constraints. This technique helps ensure that tasks are eventually completed, and your system remains resilient to failures.

By leveraging asynchronous processing and background jobs, you can offload resource-intensive tasks from your API, helping maintain its performance and responsiveness as your SaaS application scales. This approach enables you to deliver a high-quality experience to your users while efficiently managing system resources.

Stateless and idempotent API design

Designing stateless and idempotent APIs is essential when scaling a SaaS application, as it ensures that your system is more predictable, easier to manage, and less prone to errors. In this section, we will explore statelessness and idempotency and their importance in API design for scalable applications.

Stateless APIs do not maintain any client-specific state between requests, meaning each request is self-contained and independent of previous requests. Implementing stateless APIs provides several benefits, as listed here:

- **Simplifies scaling**: Stateless APIs are more straightforward to scale horizontally since you can distribute requests across multiple instances without worrying about maintaining the session state

- **Improved reliability**: Stateless APIs are more resilient to failures, as any instance can handle a request without depending on the state of other instances

- **Enhanced performance**: Stateless APIs can better utilize caching mechanisms, as responses are not dependent on the client-specific state

To design stateless APIs, consider the following practices:

- Avoid server-side sessions and, instead, use tokens (e.g., **JSON Web Token (JWT)**) to authenticate and authorize requests

- Store any required state on the client side or in external storage, such as databases or caches

An idempotent API operation, when called multiple times with the same inputs, produces the same results and side effects as if it were called once. Designing idempotent APIs ensures that your system behaves predictably and is less prone to errors due to network retries, timeouts, or other issues.

To design idempotent APIs, consider the following practices:

- Use appropriate HTTP methods, such as GET, PUT, and DELETE, which are inherently idempotent

- Implement idempotency keys or tokens for non-idempotent operations such as POST, allowing clients to safely retry requests without causing unintended side effects

- Ensure that your API's internal logic can handle repeated requests without creating duplicate records or performing unwanted actions

By designing stateless and idempotent APIs, you can build a more scalable, reliable, and predictable SaaS application. These design principles help ensure that your system can handle increased load and provide a high-quality experience for your users as your application grows.

Security and authentication at scale

As your SaaS application grows, ensuring the security and proper authentication of your API become even more crucial. In earlier chapters, we discussed building authentication into your application. Scaling your application can introduce new security challenges, and it's essential to implement robust security measures to protect your users' data and maintain their trust. In this section, we will discuss key considerations and best practices for enhancing security and authentication when scaling your API.

Using a centralized authentication and authorization system, such as OAuth 2.0 or OpenID Connect, allows you to manage user access to your API effectively. Implementing **single sign-on (SSO)** enables users to access multiple services within your application using a single set of credentials. Furthermore, utilizing an identity provider (such as Azure Active Directory) can offload the management of user identities and authentication processes, helping to ensure a secure and scalable solution.

Proper API key management is essential for maintaining the security of your API. This includes the generation, distribution, and revocation of API keys. Ensure that API keys have appropriate access levels and scopes to limit their usage to specific resources and actions. Regularly rotate API keys and encourage clients to do the same to reduce the risk of unauthorized access.

Use HTTPS for all API communications to protect data in transit and consider using technologies such as **HTTP Strict Transport Security** (**HSTS**) to enforce secure connections. Encrypt sensitive data at rest using strong encryption algorithms and key management practices. Implement proper data handling procedures to minimize the risk of data leaks or breaches.

Apply rate limiting and throttling policies for your logins to protect your API from abuse, **Denial of Service** (**DoS**) attacks, and excessive resource consumption. Customize rate limits based on factors such as user roles, API keys, or IP addresses to provide a fair and secure API experience.

Conduct regular security audits and vulnerability assessments to identify potential weaknesses in your API and infrastructure. Establish a process for addressing identified security issues and continuously improve your security posture.

By focusing on security and authentication when scaling your API, you can protect your users' data, maintain their trust, and ensure the continued success of your SaaS application. Implementing robust security measures is essential to providing a secure and reliable API experience for your growing user base.

Scaling the API of your SaaS application is a critical aspect of ensuring the overall performance, reliability, and security of your system. By addressing key areas such as stateless and idempotent API design, load balancing, versioning, rate limiting, caching, asynchronous communication, and security, you can build a robust and scalable API that can handle the demands of a growing user base.

Throughout this section, we have explored various techniques and best practices to ensure that your API can adapt to the increased demands of a successful SaaS application. By implementing these strategies, you not only enhance the performance and efficiency of your API but also ensure a consistent and secure experience for your users.

As your application continues to grow, it's essential to monitor and refine your API scaling strategies, adapting to new challenges and evolving requirements. By doing so, you will ensure the long-term success and sustainability of your SaaS application while providing a high-quality experience for your users.

Scaling the UI

Having covered database and API scaling, we will now look at scaling techniques for the UI layer. The UI is a critical component of your SaaS application, as it's the layer that users directly interact with! The user's impression of your entire application will be based on how much they enjoy (or do not enjoy) using your UI! Ensuring a smooth and responsive user experience as your application grows is essential for maintaining user satisfaction and engagement. In this section, we will discuss various techniques and best practices for scaling the UI layer, focusing on performance optimization, efficient management of static assets, and implementing effective caching strategies. Hopefully, these techniques will put a smile on your users' faces and keep them coming back to your SaaS application!

Best practices for designing scalability and performance for the UI

Good design underpins all aspects of scaling the application, including good database design and sound architectural principles on the backend. Good design on the front end is multi-faceted, though, as the design must be technically sound and also a joy for the ends users to interact with.

Designing a performant and scalable UI is tricky, though. This involves designing the UI and **user experience** (**UX**) to adapt to the growing user base and the increasing complexity of your application. By adhering to best practices, you can provide a responsive, efficient, and enjoyable experience for your users. In this section, we will explore various UI and UX best practices to help you design a scalable UI.

Keep the UI as simple and intuitive as possible to reduce the cognitive load on users. This sounds obvious and easy, but in practice, this can be incredibly challenging. Try to focus on core features, minimize visual clutter, and prioritize user workflows. A clean and straightforward UI can also help reduce the amount of processing and rendering required by the client, thus improving performance.

Ensure your application's UI adapts seamlessly to different screen sizes, resolutions, and device types. Implement responsive design techniques such as fluid grids, flexible images, and CSS media queries to create a consistent experience across various devices. This approach improves usability and helps future-proof your application as new devices and screen sizes emerge.

The UI is all that your users really see, and they will judge the performance of the entire application on the performance of the UI. Improve UI performance by optimizing rendering and reducing unnecessary re-renders. Techniques such as a virtual **Document Object Model** (**DOM**), debouncing, and throttling can help minimize the frequency of updates and the impact on performance. Additionally, consider using lighter-weight UI frameworks and libraries.

Finally, always keep accessibility in mind. As your app user base grows, the number of differently abled individuals using the app will also grow proportionately. Design your application with accessibility in mind, ensuring that it can be used by individuals with various abilities and disabilities. This broadens your user base and makes your application more user-friendly and versatile. Utilize semantic HTML, **Assistive Rich Internet Applications** (**ARIA**) roles, and keyboard navigation to enhance accessibility.

Optimizing static assets and bundling

Static assets, such as images, stylesheets, and JavaScript files, play a significant role in the performance and responsiveness of your UI. Properly optimizing and bundling these assets can lead to faster load times and a better overall UX, as well as reduce the load on your cloud resources. In this section, we will discuss several techniques for optimizing static assets and bundling them efficiently.

Minifying your CSS and JavaScript files by removing unnecessary characters, spaces, and comments can significantly reduce their size. This, in turn, reduces the time required to download and parse these files. Additionally, compressing files using algorithms such as Gzip or Brotli can further decrease file sizes, resulting in faster load times.

Optimize images to reduce their file sizes without compromising quality. Use appropriate formats (e.g., JPEG for photographs, PNG for graphics with transparency, and SVG for vector images) and ensure images are compressed to minimize their file sizes. Also, leverage responsive images and serve different image sizes based on the user's device and screen resolution.

Combine multiple CSS and JavaScript files into a single bundle to reduce the number of HTTP requests made by the client. This helps improve page load times and can be done using build tools such as webpack, Rollup, or Parcel. You can also split your bundles into smaller chunks using code-splitting techniques to load only the necessary code for a particular page or feature.

Just as in the database and API layer, we can utilize caching to optimize the UI as well. Set appropriate cache headers for your static assets to allow browsers to cache these files, reducing the need to download them again on subsequent visits. Configure cache control headers such as Cache-Control and ETag to ensure efficient caching behavior. This reduces the load on your server and improves the UX by delivering assets more quickly.

CDN serves your static assets from geographically distributed servers. This reduces latency by serving assets from a server that is closer to the user's location. CDNs also help to balance the load on your servers, improving performance and scalability.

Implement HTTP/2, which is the latest version of the HTTP protocol, to enable faster and more efficient communication between the client and the server. HTTP/2 provides benefits such as multiplexing, header compression, and server push, which can significantly improve the loading and rendering of static assets.

Optimizing static assets and bundling them efficiently can make a huge difference to the performance of the UI and also take a significant load (and therefore cost) off the cloud systems.

Implementing progressive loading and lazy loading techniques

It is very common when starting to build a UI to simply send everything that the user needs for a certain page at the initial page load. This seems like it gets all the loadings out of the way in one go and allows for the most performant UI. But, taking this approach can consume a lot of bandwidth and drive up the cost of the cloud systems. Progressive and lazy loading techniques can help to mitigate this by minimizing the amount of data and resources loaded initially, resulting in faster initial page loads and reducing the bandwidth requirements of the server/cloud.

Progressive loading involves loading content in stages, starting with lower-resolution or simplified versions and gradually replacing them with higher-quality or more detailed versions as they are required. This approach is particularly useful for images and other media, allowing users to start interacting with the content before it's fully loaded. One method to implement progressive loading is by using **low-quality image placeholders (LQIP)** or blurred thumbnails that are replaced with full-resolution images as they become available. It may be that some images never need their full resolution versions loaded, ultimately reducing the bandwidth consumed and speeding up the UI for the end user.

Lazy loading, on the other hand, defers the loading of non-critical or off-screen resources until they are needed. This technique reduces the initial payload size, resulting in faster page load times. For images and media, you can enable native lazy loading in modern browsers by using the `loading="lazy"` attribute for the `img` and `iframe` elements. If native lazy loading is not an option or you need more customization, you can also implement custom lazy loading using JavaScript libraries, such as the Intersection Observer API, which detects when elements become visible on the screen and loads them only when necessary.

In addition to images and media, lazy loading can be applied to other parts of your application, such as loading components or modules on demand. This can be particularly beneficial in large applications with numerous features or components, as it allows you to load only the necessary parts of your application when they are needed, reducing the initial load time and overall resource usage.

For example, in a Blazor WebAssembly application, you can use the built-in code splitting and lazy loading features to load specific components or entire assemblies on demand. By leveraging this technique, your application can become more modular and efficient, making it easier to scale and maintain over time.

Implementing progressive loading and lazy loading techniques in your application can significantly improve its performance, responsiveness, and overall UX. By minimizing the resources and data loaded initially and focusing on delivering only what is necessary when it's needed, you can ensure a smooth and fast experience for your users!

Leveraging caching strategies for UI components

Once again, caching is an important technique for improving the performance and responsiveness of your SaaS application's UI, particularly as it scales. By storing and reusing previously fetched or computed data, caching reduces the need for redundant requests, decreasing the load on your servers and improving the overall user experience.

One of the most effective caching strategies for UI components is client-side caching. By storing frequently used data or rendered components in the browser's cache, your application can quickly access this information without requiring additional server requests. HTML5 local storage and IndexedDB are two examples of client-side storage mechanisms that can be used for caching data.

Another caching technique involves memoization, a method for caching the results of function calls based on their input parameters. In the context of UI components, memoization can be used to cache the output of computationally expensive or frequently executed functions, reducing the need for redundant computations. Many modern UI libraries, such as Blazor, provide built-in support for memoization, making it easier to implement in your application.

When leveraging caching strategies, it's crucial to strike a balance between caching data for performance benefits and ensuring that the data remains fresh and up-to-date. To maintain data consistency, you should implement cache invalidation strategies that expire or update cached data when it is no longer valid or when the underlying data changes. Some methods for cache invalidation include setting expiration times for cached data, using versioning or timestamps to detect changes, and listening for server-side events that indicate updates to the data.

In distributed environments, such as microservices-based architectures, caching can also be implemented on the server side. Techniques such as caching API responses or using a distributed cache, such as Redis or Memcached, can help reduce the load on your backend services and improve the overall performance of your application. When implementing server-side caching, be sure to consider factors such as data consistency, cache coherence, and fault tolerance.

Caching is always a hard thing to get right, and this is no different when thinking about caching on the UI layer. It is essential to carefully plan and implement caching strategies for UI components that consider both the performance benefits and the potential complexities introduced by caching. By choosing the right caching techniques and striking a balance between performance and data freshness, you can significantly improve the UX of your SaaS application as it scales. Remember to monitor and evaluate the effectiveness of your caching strategies over time, making adjustments as necessary to ensure optimal performance and scalability.

Scaling the UI layer of your SaaS application is a crucial aspect of ensuring a smooth and responsive user experience as your application grows. By focusing on performance optimization, efficiently managing and delivering static assets, implementing progressive and lazy loading techniques, and leveraging caching strategies for UI components, you can significantly improve your application's performance and responsiveness even as it scales to accommodate more users.

As your application continues to grow, it's essential to continually monitor and refine your UI scaling strategies to ensure optimal performance and user experience. Keep in mind that adopting a data-driven approach to performance optimization, analyzing user feedback, and staying up-to-date with the latest industry best practices will help you maintain a competitive edge and deliver a high-quality experience to your users. With thoughtful planning and execution of UI scaling strategies, your SaaS application will be well equipped to handle the challenges that come with growth and expansion.

Summary

This chapter focuses on the challenges and best practices for operating a SaaS application built with Microsoft technologies at scale. The chapter is divided into four main sections: a general overview, scaling the database, scaling the API, and scaling the UI.

The first section provides a general discussion of the challenges associated with operating at scale, including infrastructure scalability, performance optimization, security and compliance, availability and uptime, and cost and resource management.

The second section covers scaling the database and includes sub-sections on partitioning, sharding, archiving, and caching. By implementing these techniques, you can ensure that your database can handle increased demand and provide reliable and performant data access for your application.

The third section covers scaling the API and includes sub-sections on load balancing, microservices, caching, and monitoring. By implementing these techniques, you can ensure that your API can handle increased demand and provide a reliable and performant data access layer for your application.

The fourth section covers scaling the UI and includes sub-sections on performance optimization, caching, load testing, UX optimization, monitoring and scaling automation, and security considerations. By implementing these techniques, you can ensure that your UI remains performant and responsive even as your user base grows and demand increases.

In summary, operating a SaaS application at scale presents several challenges, but by implementing appropriate techniques and best practices, you can ensure that your application can handle increased demand and provide a reliable and performant user experience for your customers.

We are approaching the end of our learning on SaaS applications with Microsoft technologies! In the final chapter, we will take a look at everything that we have covered and summarize our learnings!

Further reading

- 36 Things You Should Know When Designing and Scaling SaaS Software: `https://medium.com/@mikesparr/things-i-wish-i-knew-when-starting-software-programming-3508aef0b257`

- Scalability: `https://learn.microsoft.com/en-us/sql/relational-databases/in-memory-oltp/scalability?view=sql-server-ver16`

- The practical guide to API management: `https://www.softwareag.com/en_corporate/resources/api/guide/api-management.html`

- ASP.NET Core Blazor performance best practices: `https://learn.microsoft.com/en-us/aspnet/core/blazor/performance?view=aspnetcore-7.0`

Questions

1. What are the key challenges faced when scaling a SaaS application?
2. How does sharding help improve database scalability?
3. What are the differences between horizontal and vertical scaling for databases?
4. How does implementing rate limiting and throttling contribute to API scalability?
5. What is the purpose of progressive loading and lazy loading techniques in UI scaling?
6. How does caching improve performance for both UI components and backend services?

Part 5:
Concluding Thoughts

This section brings the book to a close with a single chapter that revises what we have learned and gives some pointers for how to apply your newfound knowledge!

This section has the following chapter:

- *Chapter 13, Wrapping It Up*

13

Wrapping It Up

First of all, congratulations on getting this far! We have covered *a lot* of material in this book! Building SaaS applications is not easy, and in getting this far, you have shown great dedication and perseverance. I trust that the chapters in this book have been engaging and enlightening for you. I hope that you now possess a solid understanding of the necessary tools and techniques for building enterprise SaaS applications using Microsoft technologies.

As we approach the end of this segment of your SaaS journey, it's essential to acknowledge and value the effort you have invested in this. By engaging with this material, you have invested in yourself, and the knowledge that you have gained will not only help you in building robust and scalable SaaS applications but will also open up new opportunities for your career growth and personal development.

In this chapter, we will cover three main topics that will help you consolidate your learning and prepare you for your future endeavors in the world of SaaS development. We will begin by celebrating your accomplishments and acknowledging the skills you've acquired throughout this book. Next, we'll recap the most important lessons from each chapter, emphasizing the value of these new skills in the SaaS landscape. After that, we'll discuss practical ways you can leverage your expertise in real-world projects and outline potential career advancements. Finally, we'll explore the next steps of your journey to continuous learning and growth, providing you with resources and tips to stay informed and connected in the ever-evolving SaaS industry.

Let's dive in and discover how you can make the most of your newfound skills and knowledge!

Well done!

Once again, congratulations on making it to the final chapter of this book. It has been a lot of fun writing this book, and I hope that it has been an interesting and informative journey for you, the reader, as well.

In this section, we'll take some time to acknowledge what has been achieved in working through this book.

Commitment to learning

By picking up this book, and working through each chapter, you have demonstrated a strong commitment to learning by investing your time and effort in understanding the complex concepts and putting them to use with the various Microsoft technologies and tools that have been demonstrated. This dedication to self-improvement is highly commendable and sets you apart from the masses in taking the time to learn and build your skill sets as a SaaS developer. This commitment to lifetime learning is an incredibly important and hugely valuable skill for developers to stay ahead of the curve and remain a valuable asset in the ever-changing and highly competitive world of software development.

As well as showing dedication to lifetime learning, and advancing your technical skills, you have also shown commitment to a growth mindset, which is invaluable in such a diverse, expansive, and rapidly evolving field as SaaS development. Having this mindset will help you adapt to new technologies and methodologies, and will allow you to turn challenges into opportunities for growth. As you progress in your career, this willingness to learn and adapt will be one of your greatest assets, making sure that you can always stay relevant and excel in your profession!

Mastering SaaS development skills

Showing a commitment to lifetime learning is a great thing in and of itself, but in addition to that, you have laid down some very strong foundations in SaaS development. We have looked into specific technologies, such as Docker, C# WebAPI, Blazor, SQL Server, and Entity Framework. We have also explored best practices for using these technologies, such as RESTful APIs, and have covered some SaaS-specific challenges, such as multi-tenancy, microservices, authentication, and authorization. These skills will not only make you a more versatile and effective developer, but they will also increase your ability to contribute to and lead projects in the increasingly important realm of SaaS development.

Beyond the technical skills you have mastered, you've also developed a much deeper understanding of the underlying principles that govern SaaS development. This understanding allows you to make informed decisions when architecting and implementing your solutions, ensuring that they are scalable, maintainable, and secure. As you work on future projects, you'll be able to use this knowledge to build more efficient and robust applications, which will ultimately lead to a better user experience and increased customer satisfaction. It's not just the tools and technologies that matter, but also the principles and practices that make you an exceptional SaaS developer!

Embracing industry best practices

One of the key learnings from this book is the ability to learn and implement industry best practices in your SaaS projects. By incorporating effective testing strategies, monitoring, logging, performance optimization, authentication, and authorization, you have become a better and more rounded developer beyond the specific focus of SaaS. Working with advanced concepts such as multi-tenancy and microservices will stand you in good stead for many years, and learning about CI/CD is applicable across the board in software engineering.

In an industry that is constantly evolving, staying up to date with best practices is essential for long-term success. Your ability to adapt and integrate these practices into your work demonstrates your commitment to delivering the best possible solutions for your users. As you continue to grow as a developer, keep seeking out and embracing new methodologies, tools, and techniques that can help you optimize your SaaS applications. This pursuit of excellence will not only lead to higher-quality software but also establish you as a forward-thinking developer with a keen understanding of the current and emerging trends in the SaaS landscape.

Personal and professional growth

Completing this book has undoubtedly contributed to your personal and professional growth. The skills and knowledge you've gained have enhanced your market value as a developer, and the confidence you've built will help you tackle more complex SaaS projects. Furthermore, the foundation you've established sets the stage for you to take on leadership roles or specialized positions within the industry, opening up new opportunities for career advancement. By continuing to learn and grow, you'll be well-prepared to make a meaningful impact in the world of SaaS development.

However, the personal and professional growth you've achieved by completing this book is only the beginning of your journey! As you continue to learn, refine your skills, and take on new challenges, you'll find yourself growing not just as a developer, but as a leader and a mentor to others. Embrace this growth and actively seek out opportunities to share your knowledge, collaborate with peers, and make a difference in the software development community. By doing so, you'll not only elevate your career but also contribute to the overall progress and innovation in the world of SaaS development.

What have you learned?

Throughout this book, you have gained valuable knowledge and insights into the world of SaaS development. Let's take a moment to revisit the most important concepts and skills you've acquired.

Understanding SaaS was the starting point of your journey. By grasping the fundamentals of SaaS, its benefits, and its architecture, you have developed a strong foundation that has prepared you to navigate the complexities of SaaS development. You learned how SaaS applications can provide cost-effective, scalable, and accessible solutions for businesses of all sizes, making them an attractive choice in the modern software landscape.

Recognizing the significance of SaaS in the contemporary software ecosystem has allowed you to appreciate the growing demand for skilled SaaS developers. As more organizations adopt cloud-based services and subscription models, your understanding of the core principles of SaaS will enable you to make informed decisions when building and deploying your applications, ensuring that you can deliver the best possible solutions to meet the needs of today's users.

As you progressed through this book, you delved deeper into essential concepts such as multi-tenancy and scalability, which play a crucial role in the success of SaaS applications. You comprehended the concept of multi-tenancy, enabling multiple clients to use a single application instance while maintaining data isolation. Understanding the importance of multi-tenancy in SaaS applications has provided you with valuable insights into designing and architecting solutions that cater to the unique requirements of diverse clients and industries.

You've learned strategies for building scalable and data-rich applications, ensuring that your SaaS solutions can grow alongside the needs of your users. In doing so, you've become familiar with databases and Entity Framework, an essential tool that simplifies data access and allows you to work more efficiently with databases. Additionally, you've explored the importance of microservices, an architectural pattern that promotes the development of modular, independently deployable services. By embracing microservices, you can create more maintainable, scalable, and resilient SaaS applications, providing a better experience for your users while simplifying the ongoing management of your software.

As you ventured into the realm of frontend development for SaaS applications, you discovered the importance of creating intuitive and user-friendly interfaces. A well-designed user interface not only enhances the user experience but also contributes to the overall success and adoption of your SaaS solution. In this context, you learned about Blazor, a powerful framework for building interactive web applications using C#. By leveraging Blazor, you can create a seamless and cohesive development experience, using the same language and tools across both the frontend and backend.

In addition to frontend development, you delved into the critical aspects of authentication and authorization. Ensuring the security of your user's data and maintaining appropriate access controls are paramount in the world of SaaS. You learned about various authentication mechanisms, such as OAuth and OpenID Connect, which enable secure sign-in processes and help protect user data. By implementing strong authorization strategies, you can ensure that users have access only to the resources and actions that are relevant to their roles and permissions.

By mastering these concepts and tools, you have acquired the necessary skills to create robust, secure, and visually appealing SaaS applications that meet the demands of your users. Combining these frontend development techniques with your knowledge of backend technologies, you are now well-equipped to build and deploy comprehensive SaaS solutions that cater to a wide range of customers and industries.

Throughout this book, you have also gained insights into crucial software engineering practices that extend beyond the specific realm of SaaS development. These practices are integral to delivering high-quality software and can be applied across various types of projects and domains.

Testing is one such essential practice, and you have learned the importance of thorough testing across the entire stack to ensure the reliability and correctness of your software. By adopting various testing strategies, such as unit, integration, and end-to-end tests, you can validate the functionality of your applications and identify issues early in the development process.

Monitoring and logging are vital components of maintaining and troubleshooting software. By incorporating effective monitoring and logging solutions into your SaaS applications, you can quickly identify and address performance bottlenecks, errors, and other potential problems. These techniques enable you to proactively manage your applications, minimize downtime, and provide a consistent, high-quality experience for your users.

You also explored the concepts of **Continuous Integration and Continuous Deployment (CI/CD)**, which promote the "Release Often, Release Early" approach. CI/CD pipelines automate the process of building, testing, and deploying your applications, reducing manual intervention and improving the overall quality and velocity of software releases.

Lastly, you learned about the challenges of operating at scale, as SaaS applications often need to serve a growing number of users and handle increasing amounts of data. By understanding scalability principles and implementing strategies to manage growing pains, you can ensure that your SaaS solutions remain performant, resilient, and reliable, even as they expand to meet the needs of a larger user base.

By mastering these general software engineering practices, you have become a more versatile and effective developer, capable of delivering high-quality, scalable, and maintainable software solutions that meet the demands of today's users.

How can you use these skills?

Equipped with a wide range of skills and knowledge in SaaS development and general software engineering practices, you are now prepared to apply these capabilities in various contexts. Whether you are seeking to improve your current projects, explore new opportunities, or make a difference in the development community, these skills open up numerous possibilities for you to make a significant impact in the world of SaaS and beyond. In this section, we will discuss various ways to utilize and capitalize on the expertise you have acquired throughout this book.

Apply your knowledge to your current job or projects

This strikes me as the most obvious and easiest way for you to start putting what you learned in this book into practice. It would, of course, be ideal if your current employer was working on a SaaS application, and you could now confidently take on a strong technical role in developing that application! Of course, that may not be the case. However, many of the skills that you learned in this book can be applied to any software engineering project. You could perhaps start to build out some additional automated testing using the skills you learned from *Chapter 9*, or you could impress your team by building or improving the CI/CD pipelines we learned about in *Chapter 11*.

More generally, you may now be able to improve some of the practices in place in your team. Optimizing the processes around the actual work of developing software is a skill in and of itself, and the learnings you have taken from this book should serve you well in that endeavor, should you choose to try.

I trust that the insights shared in this book will equip you with the necessary skills to undertake and execute a SaaS project successfully. I would also hope that enough generally good advice has been covered that you can start to put at least some of these skills into practice almost immediately and start to deepen your knowledge and understanding through practical application of the topics that we have covered.

Freelance or consulting opportunities

The field of software engineering is vast and varied, and while you may immediately start to use your newfound knowledge in your current place of employment, that will not be the case for everyone!

Another avenue to explore with your newfound SaaS development expertise is the world of freelancing or consulting. Many businesses are seeking skilled professionals to help them develop, maintain, or improve their cloud-based solutions. As a freelance developer or consultant, you can offer your services to clients on a project-by-project basis or provide ongoing support, depending on their needs.

By helping businesses transition to cloud-based solutions, you can play a crucial role in their digital transformation journey. Your knowledge of SaaS development, combined with your understanding of industry best practices and software engineering principles, can enable you to provide valuable guidance to clients. You can help them optimize their existing applications, identify opportunities for innovation, and streamline their development processes, all while delivering tangible results that improve their bottom line.

As a freelancer or consultant, you also have the opportunity to build a diverse portfolio, work with a variety of clients, and tackle new challenges. This can be an excellent way to further expand your skill set, gain exposure to different industries and technologies, and make a meaningful impact on businesses and their customers.

Build your own SaaS product or start-up

Freelancing is a great way to build up your skills and contribute to a wide variety of projects. However, you may also want to consider building out your idea. Launching a start-up is undoubtedly a challenging approach to take! Start-ups require a significant amount of work, and the skill set needed goes way beyond the technical ability to develop a SaaS application. However, the potential rewards of venturing down this road are also massive.

If you do decide to go down this route, begin by identifying a niche or problem that your SaaS solution could address. This could be a pain point in a specific industry or a more general challenge faced by businesses and users alike. By focusing on a unique and valuable proposition, you can create a product that stands out in the market. It is very good, and very common advice, to "scratch your own itch," by which I mean to build something that solves a problem that you have. In this way, you can be your own customer-number-one, and use your insights into the problem domain to make a great product for your subsequent customers!

As the founder of a SaaS start-up, you'll need to wear many hats. Your responsibilities will not only encompass the technical aspects of creating the application but also involve devising a business model, pricing strategy, and go-to-market plan. Additionally, you'll need to oversee product development, marketing, sales, and customer success. These are skills that are not always associated with software engineers! Don't be afraid to plug these gaps by bringing others along on your start-up journey.

Growing your own SaaS business can be both challenging and rewarding. As you navigate the ups and downs of entrepreneurship, you'll have the opportunity to learn from your experiences, adapt to changing market conditions, and build a team that shares your vision. By applying the skills you've gained throughout this book, you can create a successful SaaS product that not only solves a critical problem but also generates sustainable revenue and growth.

Contribute to open source projects

Freelancing and launching a start-up are both great ways to start using your SaaS skills, but both are undoubtedly a lot of work. Many of us live busy lives and simply do not have the time to devote to such an undertaking. Another route to working with a SaaS application is to contribute to an open source project.

Contributing to open source projects allows you to enhance your skills while giving back to the community. By sharing your knowledge and expertise, you can help improve existing projects and drive innovation in the software industry. Open source projects often have diverse and welcoming communities, where you can find mentorship, support, and camaraderie.

Collaborating with other developers in open source projects also offers an invaluable opportunity to learn from their expertise. You can gain exposure to new technologies, programming languages, or frameworks, as well as different coding styles and best practices. This can significantly broaden your horizons and further refine your development skills.

Getting involved in open source projects also has other benefits, such as building your portfolio, expanding your professional network, and potentially even leading to job opportunities. By dedicating some of your time and energy to contributing to these projects, you can have a lasting impact on the software development community while simultaneously advancing your career.

Stay up to date with industry trends and best practices

In simply picking up and reading this book, you have demonstrated your dedication to maintaining your skill set and staying abreast of the latest technological advancements and industry best practices. In the field of software engineering, this pursuit is a lifelong endeavor that does not conclude as we approach the end of this book. There is always more to learn, and I encourage you to continue your journey of learning and growth beyond these pages. One of the best ways to stay current with industry trends and best practices is to attend conferences, workshops, and webinars. These events provide opportunities to learn from experts, discover new tools and technologies, and network with other professionals in the field. By actively participating in these gatherings, you can ensure that your knowledge and skills remain relevant in the ever-evolving world of software development. Often, these conferences are looking for speakers as well. You can challenge yourself by volunteering to talk at a conference!

Additionally, make it a habit to follow industry leaders, blogs, and social media channels. This will help you stay informed about emerging trends, innovative solutions, and the latest insights from thought leaders in the SaaS and software engineering space.

Staying up to date with the latest developments in the industry is important for any software professional. By dedicating time and effort to continuous learning and growth, you can keep your skills sharp and be better prepared to adapt to the changing landscape of software engineering and SaaS development.

Mentor and teach others

Finally, a very rewarding way to keep your skills sharp can be to mentor and teach others. Sharing your knowledge and expertise not only helps others grow in their careers but also reinforces your understanding of concepts and practices. Teaching can be an excellent way to reflect on your own experiences, identify areas for improvement, and stay connected to the fundamentals of SaaS development and software engineering.

Consider sharing your knowledge with colleagues at work or in online communities, such as forums, social media groups, or platforms such as Stack Overflow. By offering guidance, answering questions, or providing feedback, you can contribute to the collective growth of the development community and establish yourself as a trusted expert in your field.

Additionally, you can offer workshops and training sessions or develop educational content, such as blog posts, articles, or video tutorials. This not only helps others learn from your experiences but also allows you to hone your communication and presentation skills, which are invaluable in any professional setting.

Mentoring and teaching others can be a fulfilling and mutually beneficial endeavor. By sharing your expertise and supporting the growth of others, you can have a lasting impact in the world of software engineering while also keeping your skills sharp and up to date.

What to do next

Now that you have finished this book, and have started to look toward the next steps in your journey as a SaaS developer, let's consider some of the specific ways you can expand your skills, expertise, and professional network.

As developers, the first thing that we must do is always stay technically sharp. It is always a good idea to deepen your expertise in specific technologies or areas, such as C#, Entity Framework, or Blazor. This will allow you to tackle more complex projects and challenges, and potentially even contribute to open source projects or create educational content. Additionally, you may want to explore alternative databases, architectures, or cloud platforms to expand your knowledge and diversify your skill set.

Alongside strengthening your technical expertise, it's important to explore related disciplines, such as DevOps, data science, or machine learning. Gaining proficiency in these areas and other programming languages or frameworks will make you a more versatile developer, opening up new opportunities and challenges. Obtaining relevant certifications, such as the **Microsoft Certified Solutions Developer (MCSD)**, Azure Developer, or Azure Solutions Architect certifications, can further enhance your credibility and marketability in the field.

Networking with other professionals is essential for staying connected to the industry and learning from the experiences of your peers. Engaging in local meetups, user groups, or online communities, as well as attending industry events, conferences, and workshops, can help you deepen your expertise while also enabling you to share your knowledge and experiences with the wider community. This might involve creating articles, blog posts, or video content, which not only establishes you as an expert in SaaS development but also benefits others in the field.

Lastly, setting personal and professional goals is crucial for guiding your career path and maintaining a sense of direction. By identifying short-term and long-term objectives, you can continuously assess your progress and adjust your plans as necessary. By actively exploring these opportunities and remaining committed to ongoing growth, you can ensure your success in the world of SaaS development and beyond.

Summary

This chapter has served as a wrap-up of all that we have covered in this book and a look at what may be in front of you now that you are equipped to become a SaaS developer! We looked at the many facets of developing SaaS applications, including multi-tenancy, microservices, UI design, testing, CI/CD, and more! We also looked at the many paths you can take to continue expanding your skills, expertise, and professional network. From deepening your expertise in specific technologies, exploring related disciplines, and obtaining relevant certifications to networking with other professionals, sharing your knowledge, and setting personal and professional goals, there are numerous opportunities for you to pursue.

As you embark on this exciting journey, remember to stay curious, open to new challenges, and adaptable to the ever-evolving world of software development. The knowledge and experience you have gained from this book will serve as a strong foundation for your future projects, whether they involve freelance work, contributing to open source projects, launching a start-up, or mentoring others.

There is nothing left to be said! I would like to thank you personally for picking up this book and taking the time to read through it to the very end. I hope that this has been an interesting and informative read. I also very much hope that this will serve as the start of many projects that you can now undertake in the world of SaaS development.

Good luck in all of your future endeavors!

Packtpub.com

Subscribe to our online digital library for full access to over 7,000 books and videos, as well as industry leading tools to help you plan your personal development and advance your career. For more information, please visit our website.

Why subscribe?

- Spend less time learning and more time coding with practical eBooks and Videos from over 4,000 industry professionals

- Improve your learning with Skill Plans built especially for you

- Get a free eBook or video every month

- Fully searchable for easy access to vital information

- Copy and paste, print, and bookmark content

Did you know that Packt offers eBook versions of every book published, with PDF and ePub files available? You can upgrade to the eBook version at packtpub.com and as a print book customer, you are entitled to a discount on the eBook copy. Get in touch with us at customercare@packtpub.com for more details.

At www.packtpub.com, you can also read a collection of free technical articles, sign up for a range of free newsletters, and receive exclusive discounts and offers on Packt books and eBooks.

Other Books You May Enjoy

If you enjoyed this book, you may be interested in these other books by Packt:

High-Performance Programming in C# and .NET

Jason Alls

ISBN: 9781800564718

- Use correct types and collections to enhance application performance
- Profile, benchmark, and identify performance issues with the codebase
- Explore how to best perform queries on LINQ to improve an application's performance
- Effectively utilize a number of CPUs and cores through asynchronous programming
- Build responsive user interfaces with WinForms, WPF, MAUI, and WinUI
- Benchmark ADO.NET, Entity Framework Core, and Dapper for data access
- Implement CQRS and event sourcing and build and deploy microservices

Parallel Programming and Concurrency with C# 10 and .NET 6

Alvin Ashcraft

ISBN: 9781803243672

- Prevent deadlocks and race conditions with managed threading
- Update Windows app UIs without causing exceptions
- Explore best practices for introducing asynchronous constructs to existing code
- Avoid pitfalls when introducing parallelism to your code
- Implement the producer-consumer pattern with Dataflow blocks
- Enforce data sorting when processing data in parallel and safely merge data from multiple sources
- Use concurrent collections that help synchronize data across threads
- Debug an everyday parallel app with the Parallel Stacks and Parallel Tasks windows

Packt is searching for authors like you

If you're interested in becoming an author for Packt, please visit `authors.packtpub.com` and apply today. We have worked with thousands of developers and tech professionals, just like you, to help them share their insight with the global tech community. You can make a general application, apply for a specific hot topic that we are recruiting an author for, or submit your own idea.

Share Your Thoughts

Now you've finished *Building Modern SaaS Applications with C# and .Net*, we'd love to hear your thoughts! Scan the QR code below to go straight to the Amazon review page for this book and share your feedback or leave a review on the site that you purchased it from.

`https://packt.link/r/1-804-61087-9`

Your review is important to us and the tech community and will help us make sure we're delivering excellent quality content.

Download a free PDF copy of this book

Thanks for purchasing this book!

Do you like to read on the go but are unable to carry your print books everywhere?

Is your eBook purchase not compatible with the device of your choice?

Don't worry, now with every Packt book you get a DRM-free PDF version of that book at no cost.

Read anywhere, any place, on any device. Search, copy, and paste code from your favorite technical books directly into your application.

The perks don't stop there, you can get exclusive access to discounts, newsletters, and great free content in your inbox daily

Follow these simple steps to get the benefits:

1. Scan the QR code or visit the link below

https://packt.link/free-ebook/9781804610879

2. Submit your proof of purchase
3. That's it! We'll send your free PDF and other benefits to your email directly

Index